D0066117

THE ICE FINDERS

LOUIS AGASSIZ

ELISHA KENT KANE

CHARLES LYELL

Selected other titles by Edmund Blair Bolles

AS EDITOR:

Galileo's Commandment: An Anthology of Great Science Writing

AS AUTHOR:

A Second Way of Knowing: The Riddle of Human Perception

Remembering and Forgetting: Inquiries into the Nature of Memory

So Much to Say

THE ICE FINDERS

Edmund Blair Bolles

How a Poet, a Professor, and a Politician Discovered the Ice Age

COUNTERPOINT WASHINGTON, D.C.

Maps by Pascal Jalabert

Library of Congress Cataloging-in-Publication Data
Bolles, Edmund Blair, 1942–
The ice finders : how a poet, a professor, and a politician discovered the Ice Age / by Edmund Blair Bolles.
p. cm.
ISBN 1-58243-030-6 (hc)
1. Glacial epoch. 2. Agassiz, Louis, 1807–1873. 3. Lyell, Charles, Sir, 1797–1875. 4. Kane, Elisha Kent, 1820–1857. I. Title.
QE697.B725 1999
551.7'92—dc21 99-35799
CIP

Printed in the United States of America on acid-free paper that meets the American National Standards Institute Z39-48 Standard

COUNTERPOINT
P.O. Box 65793
Washington, DC 20035-5793

Counterpoint is a member of the Perseus Books Group

10 9 8 7 6 5 4 3 2 1

Sisters and brothers sure were a good idea.

Here's to Zoé, DeVallon, and Harry.

Contents

Part 1

Ignorant, Ambitious Men

The most famous law of mechanical intelligence says: *Garbage in, garbage out.* If a machine has no information and no rules for deducing the information, it stays pig-ignorant. As someone who has worked with and developed many databases in the corporate world, I can testify that the first job in launching a new system is making sure the initial data is right. The second job is developing methods that will make sure the data stays right. Companies often try to save money by skimping on the second or even the first of these requirements, and then, *look out!* Why, nineteen years after her death, am I still receiving junk mail addressed to my grandmother? Recently I was even offered a preapproved credit card in her name. True, my grandmother worked hard to maintain a perfect credit rating, but some group has let its database go to seed.

A strange twist, however, is that this basic law of machine intelligence does not always hold true for people. There have been people like the alchemists who were wrong from start to finish, but more often people do learn unsuspected things, pulling knowledge, like rabbits, from empty hats. How can human intelligence do so much better than machines?

The Ice Age is just such new knowledge. There was a time when nobody knew that much of the northern hemisphere, and parts of the southern one too, had been covered under monstrous ice caps a mile and more thick. Today anybody who stayed awake at school knows about that Ice Age. Something must have happened to make us wiser. It is not just that we got a new fact or two. Our whole imagination about the earth and its ice has changed. Anyone reading this book already enjoys some knowledge of the conquering ice—of great ice, ice a mile thick, ice that flows like the Nile and sweeps castle-sized boulders along with it, ice that carves mountains into valleys. When we worry that global warming might melt the polar ice caps and drown Florida, we are picturing relationships that were once unknown to anybody alive.

For most of history, most people never saw, never heard of the great ice. Presumably Erik the Red and a few other Vikings who made it to Greenland had some idea. Eskimos knew. Penguins knew. Most everybody else had not a hint of it, and, not knowing, they got things exactly backward. In the 1820s, Robert Jameson, a naturalist at the University of Edinburgh and editor of an important journal, was telling his students that the greatest ice in the world was found in the temperate climates. As people travel north toward the poles or south to the equator, Jameson said, glaciers diminish in size and finally disappear altogether. That little doctrine not only denies the snows atop Mount Kilimanjaro, but it also says there can be no ice at the poles. Garbage in, garbage out.

The British began serious Arctic exploration at about the same time that Jameson wrote, so it might seem as though a few treks into the north would have been enough to bring home the truth about ice. Yet thirty years later, in the 1850s, geographers still argued over whether the North Pole was frozen solid. A recurring scene in the Arctic explorations of that period featured a shipload of wretched, underdressed sailors. They shivered while their gums, rotten with scurvy, bled. Their vessel was frozen in ice, but their eyes aimed northward, focused on some hazy break in the floe, as the most learned and optimistic among them said, "There, that way lies the route to the iceless, polar sea."

Hunh? the modern reader wants to ask. Were they looking for mermaids too?

Granted, they had not yet reached the Pole and could not be dogmatic about its nature, but surely common sense tells us that travelers get colder and the sea turns icier as we go farther north. Common sense, however, also tells you the earth is flat. Scientists and determined leaders do not think much of common sense, and the nineteenth century was an age of science and determination. All assumptions had to be reexamined. Optimists noted that even the most northerly explorers had seen birds migrating still farther north, and whalers between Alaska and Siberia had once bagged an animal that bore a harpoon marked by a ship in Baffin Bay, between Greenland and Canada.

Even so, anybody who sailed the Arctic waters (whaling crews mostly) scoffed at the fantasy of an Open Polar Sea. It took men with big ideas and no sea legs to think otherwise, men like the American naval officer Matthew Fontaine Maury. Maury is still remembered as the father of oceanography and as the creator of unusually accurate charts of the seas' winds and currents. He was the very opposite of a fool, but he was also a dryland sailor who had been injured early in his career and became tied to a desk. Trapped like that, stuck in his office as firmly as a ship locked in a frozen sea, he might have pictured the helplessness of icebound voyagers. Instead, just as he could imagine a more active self who had gotten beyond his desk, he imagined those icebound sailors traveling beyond their prisons to flowing waters.

In 1850, Maury added his theory of an Open Polar Sea to the written orders given an expedition heading north. An American party was sailing beyond the Arctic Circle in search of Sir John Franklin, a British explorer who had disappeared some four years earlier. Maury's orders argued that beyond the "icy barrier" there might be a "sea free from ice." The orders even included instructions for what to do if it turned out to be easier to sail through the iceless sea than to return to the Atlantic. ("Put yourself in communication with any of the United States naval forces or officers of the government serving in the waters of the Pacific or in China.")

Seventeen months after receiving those orders the expedition was back from the Arctic, and its leader, Edwin De Haven, offered his report. His team had not found Franklin and had not reached an iceless sea, but De Haven had seen "a wide channel tending to the westward. A dark, misty looking cloud which hung over it...was indicative of much open water in that direction." The channel was christened Maury, "after the distinguished gentleman...whose theory with regard to an open sea to the north is likely to be realized through this channel."

A year later (1852) another explorer, still looking for Franklin, made it all the way to the top of Baffin Bay and entered an unexplored water called Smith Sound. A gale immediately blew the ship back into the bay, but the sound had appeared to be free of ice for at least 30 miles. The ship returned to England with the news that maybe the route to the Open Polar Sea pointed above Greenland, and the next spring still another ship set off to have a look. Its captain was Elisha Kent Kane, who had been ship's surgeon on De Haven's voyage and who had seen the frosty mist over Maury Channel. Before sailing on this new venture, Kane told an audience of learned New Yorkers, "The circumpolar ice...is an annulus, a ring surrounding an area of open water....My plan of search is...[to] hasten to Smith's Sound, forcing our vessel to the utmost navigable point....We can push forward our provisions by sledge and launch. Thus equipped, we follow the trend of the ocean, seeking the *open sea.* Once there, if such a reward awaits us, we launch our little boats, and, bidding God speed us, embark upon its waters."

Kane was a romantic, an adventurer in love with big ideas, but more sober types filled his audience. He gave his lecture at a regular meeting of the American Geographical and Statistical Society. The society's members were part of the nineteenth century's continuing effort to replace ignorance with truth. They knew that immense ice floes and bergs dominated the Arctic waters, and yet the facts of great ice were too strange for their imaginations to accept. They still dreamt in terms of winter's frozen lakes and ponds—of ice that gets thinner as you push out from shore and that, if the water is big enough, often has an open area in its middle. Picturing what they already knew, they could not see that conquering ice belongs to a different species altogether.

LONDON, JULY 1830...

The nineteenth century's most influential geologist also began his career with a display of total ignorance about great ice. He was Charles Lyell, a Scottish geologist who hurled himself into the public eye by publishing a popular book that listed every phenomenon thought capable of changing the earth's surface. Somehow Lyell's book managed to avoid mention of glacier effects altogether, despite his having grown up surrounded by major evidences of glacier activity. Within a few miles of his family's home in the Scottish Lowlands stood a "moraine," a pile of rubble that had been pushed up by a moving ice wall. Lyell saw it but did not understand. The valley walls near the estate showed long scratches and streaks of polished rock, cliff sides that had been sandpapered smooth, sometimes almost to the gloss of a marble floor, by a passing glacier. Lyell saw them and supposed that polished rocks were a commonplace of nature. Scattered about Scotland also are many enormous boulders—some especially mountainous ones surround the city of Edinburgh—that have been sprinkled over the land in some seemingly random way. Lyell saw them and could see they had been carried there from somewhere else, but he seems never to have imagined that they might once have ridden on top of a glacier.

Lyell's book *The Principles of Geology* had a special point to make: The world we see today can be explained by the geological actions we see today. He rejected both traditional explanations for the landscape, like Noah's universal flood, and speculative catastrophes, like the near-instantaneous rise of a mountain group that some geologists imagined. If Lyell was going to explain every dip and bump in the face of the earth by citing only contemporary causes at their contemporary strengths, he needed a full list of available explanations. So his book detailed the effects of lava flows, earthquakes, geysers, tidal motions, soil erosion, the drying up of swamps, and even the transportation of rocks by icebergs. Yet, for all his knowledge and ingenuity, Lyell said nothing about the effects of glaciers upon the earth.

In 1830, the best example of Ice Age quality ice near Europe lay in Greenland, an enormous island covered by one continuous sheet of ice. Glaciers oozed out on all sides, especially along its western coast where ice still breaks off in huge chunks that float down through Baffin Bay and interfere with ocean shipping. Whaling ships had been sailing Baffin Bay for centuries and were well skilled at surviving the massive ice that broke off from the Greenland sheet. Greenland was one of the most spectacular geological sites in the world, but Lyell knew almost nothing of it. He wrote about icebergs in his book and briefly mentioned Greenland, but he got its geography confused and said Baffin Bay was on Greenland's east coast.

As far as the Ice Age was concerned, Lyell's book was another case of garbage in. There were perhaps no more than five scholars scattered across the earth in Scotland, Scandinavia, Switzerland, and Germany who might have noticed the omission. Two generations earlier a great and admired scientist, Horace Bénédict de Saussure, had studied the Alps and given a good description of glaciers, yet Lyell could ignore his work and get away with it. Partly this was because Saussure never accounted for anything in Switzerland as the remains of an ancient glacier. For most naturalists and educated readers, Lyell's book was a triumph of facts and ingenious explanations. Established geologists disagreed with this or that point, but all were impressed by how much Lyell knew and by how he could tie together facts from different sides of the globe.

Even more impressed was the generation of naturalists just coming of age. Charles Darwin, for example, was about to enter his fourth year of studies at Cambridge. Years later Darwin wrote that Lyell's *Principles* "altered the whole tone of one's mind, and therefore, when seeing a thing never seen by Lyell, one saw it partially through his eyes." It was a great compliment, but one that meant that a younger generation expected nothing from glaciers.

PARIS, MAY 8, 1832...

The Baron Georges Cuvier did not think much of Lyell. He had met the young Englishman, of course. Lyell had paid his respects to France's leading zoologist when he came to Paris. Cuvier's salon was the scientific equivalent of France's most fashionable aristocratic circles, and Lyell was not so foolish as to stay away. But Lyell was a theorizer, and Cuvier had no use, he said, for "principles." He wanted facts. With facts, Cuvier had invented paleontology and had turned anatomy into a tool of discovery. He was, in the spring of 1832, giving a series of lectures on the history of science. The moral was always the same: Forget theory and go for the facts.

The lecture scheduled for May 8 was greeted with an extra dose of anticipation, if only because it had been postponed before. Paris that spring was suffering from a cholera epidemic, and large gatherings were a health risk. With the epidemic apparently on the wane, the event was rescheduled.

The impression Cuvier made as he appeared on stage would surely startle a modern audience. He was sixty-three, fat, and dressed like a French aristocrat in clothes that did not camouflage his portliness. Yet his knowledge was prodigious, his speech energetic, and his passion as intense as Romeo's. His love of facts did not make him boring. He lined up his data like pearls on a string, so that each added luster to its neighbor. His audience of students, other naturalists, and stylish aristocrats eager to be seen by other stylish people was taking in a performance as well as a lecture.

Besides mastering nature's facts, the baron had sound political and social skills. He had survived France's tumultuous history in a manner that suggested a very alert guardian angel. He began to be noticed during the early days of the French republic, was honored by Napoleon as a star of the revolutionary empire, was honored again when the Bourbon monarchy was restored, and now, after yet another revolution, he was still a hero of the city. His only challenger for the title of Europe's greatest naturalist was a German baron, Alexander von Humboldt, who had taken a seat in the audience to hear Cuvier's consideration of that day's topic, "*Naturphilosophie*, evolution and unity." Twenty-seven years before Darwin's big book, Cuvier proposed to consider German and French theories about the transformation of species.

The major philosophical difference between Darwin and earlier evolutionists was that pre-Darwinians saw evolution as a form of progress. Giraffes grew long necks and became better giraffes for it. The Germans especially saw nature as striving toward ideal forms. Cuvier scoffed at these ideas, and as he lectured, he bellowed his scorn. His own observations had shown him that each species was already perfectly suited to its condition and did not need to improve. More importantly, however, he rejected these theories because they were theories and, therefore, an impediment to scientific progress. Over the centuries many theories had come and gone, leaving nothing of benefit, but a fact was a fact forever.

Some in the audience agreed, while others probably had no idea what he was talking about. Cuvier was bringing republicanism into natural science by denouncing the very old custom of ranking life-forms into a hierarchy. Ever since the ancients, philosophers had spoken of a "Great Chain of Being" that ordered all living things into a line that ran from the lowest form of pond scum on up to humans, the angels, and, at the very top, God Almighty. Language from that image is still with us today. Talk of food chains and missing links hearken back to that old image of a Chain of Being. When we speak of "lower life-forms" we are taking the old Chain of Being's assumption that we rank higher than the plants and animals.

Cuvier thought this ranking was absurd. Why should a spider rank lower than a fish? Each had its role and position, and each played it admirably. "Each being," he told that afternoon's audience, "contains in itself, in an infinite variety and in an admirable predisposition, all that is necessary for it. Each being is perfect and viable, according to its order, its species, and its individuality." At least one aristocrat in the audience reported taking this remark to have the highest religious meaning, and he nearly swooned for the nobility of its sentiment. Cuvier was a Protestant and a believer, but that lecture-hall passage sounds perilously close to saying that each lily is so well suited to its field that, once created, it needs no further divine attention to thrive. It is the zoological equivalent of Isaac Newton's universe, which, once set in motion, needs no further intervention from a Creator.

Traditionally, nothing in nature was thought to be as independent as Cuvier described species. The Great Chain of Being had been a real chain, held together by its links, and the whole order of nature required the continued support of each part of the chain. As Alexander Pope put it:

From Nature's chain, whatever link you strike,
Ten or ten thousandth, breaks the chain alike.

Cuvier had accumulated the facts needed to demolish Pope's image. He proved that the supposedly unbreakable chain had been broken many times because creatures that had once thrived were now extinct. The old theory had had to give way before the facts that species of elephant, rhinoceros, giant sloths, and many other animals had ceased to exist. They were known through their fossils, but not through any living creature. Cuvier's masterpiece proof of extinction had studied the mammoth elephant. He analyzed the creature's anatomy, establishing that it differed from the both Asian and African versions of living elephants. The mammoth was a distinct elephant species, too big to have gone unnoticed if it were still alive. If they were in South America, Humboldt would have learned of them. In North America? Lewis and Clark or one of the many fur traders would have gained some hint of their existence. The facts proved that extinction was real.

As Cuvier lectured, Humboldt grumbled to his companion that Cuvier was too unimaginative. Cuvier, of course, would have denied that imagination had any role in science. The companion was startled by the gibe, but Humboldt loved to snip at Cuvier. When Lyell had visited Paris, Humboldt surprised him by giving him all the nasty gossip. Lyell told his father, "We must not forget that Baron Humboldt and he [Cuvier] are the two great rivals in science."

Although Cuvier's evidence had ended the debate over the extinction of species, a few biologists were still scandalized by the idea. What was the Creator doing? Making and destroying living forms for sport? Most scandalized of all was Cuvier's colleague in Paris, Jean-Baptiste Lamarck, who found an ingenious way to deny extinction. Instead of proposing that ancient species were still to be found in some as-yet-undiscovered valley, Lamarck argued that ancient species lived today, transformed into new species. The plant and animal species found today, said Lamarck, are the direct descendants of the lost species found only in fossil form. Cuvier despised this attempt to restore the Great Chain of Being to biology. The rivalry between the two men became so great that, upon Lamarck's death, Cuvier wrote a "eulogy" that was so insulting to Lamarck it was not published for years.

Humboldt's grumbling during the lecture did have a point. Cuvier had fought against theory for so long that he failed to notice that theory is everywhere. Even facts become theories the moment you assign them a meaning. Extinction was a fact, but Cuvier had given it the role of a theory when he used it to argue, in principle, against evolution.

There seemed to be a contradiction in Cuvier's lecture. If each species is perfect, why do species become extinct? But

Cuvier had an answer: Periodic local catastrophes do them in. In the case of mammoths, Cuvier suggested that some kind of cold snap had wiped them out. He had once examined the puzzling case of a mammoth that had been found frozen in Siberia. Cuvier suggested it had been frozen in its tracks by a sudden drop in the temperature. Frost will destroy wheat fields and orange trees. Perhaps a truly great frost could stop an animal, even a huge one, right where it stood, killing it as easily as nature breaks a stalk of corn.

More often, Cuvier said, the catastrophe had been local flooding. The geological records showed there were many sudden breaks in life-forms and terrain. These local revolutions had produced many worlds as old species were wiped out and replaced by new creations. This kind of thinking was what Lyell protested. Catastrophes seemed too arbitrary and easy. Especially maddening was Cuvier's pretension that his theory of revolutions was no theory at all, just a fact.

When the lecture was over and the audience made its way out, it was already clear that the talk had been a great success, the pinnacle of his series, people said later. It was surely an impressive performance, although perhaps memories were influenced by the fact that the next day the baron fell sick and took to his bed. Four days later he was dead of the cholera.

With Cuvier's death, an era seemed to have ended, but his influence lived on more energetically than anyone could have anticipated. The colleague who sat beside Humboldt at the lecture had been a very young, very able Swiss naturalist named Louis Agassiz. He, along with Elisha Kane and Charles Lyell, is one of the three heroes of this history. Together they changed our understanding of the earth's history, climate, and the role of ice on earth. They seem a perfect case for observing the difference between human understanding and machine intelligence. They certainly started out with full loads of garbage in. Besides Kane's denial of polar ice and Lyell's ignorance of glacier effects, Agassiz began his career as a loyal Cuvierian. He even believed in a catastrophic refrigeration so powerful that it could stop an elephant.

NEAR ETAH VILLAGE,
GREENLAND, JANUARY 29, 1855...

Ignorance of great ice is one thing for scholars in Paris and London but quite another for Arctic adventurers. Elisha Kent Kane's role in this history is that of a poet. He is the one who made the Ice Age imaginable. You cannot conceive of how elephants ended up in a block of ice? You will if you read of Kane's troubles. He raised the money and sailed north beyond Baffin Bay and into Smith Sound. There he became trapped on Greenland's north coast and, by the beginning of 1855, had been stuck there for seventeen months.

Both he and his crew were starving, reduced to living on the poisonous liver of a polar bear. In desperation Kane and his Eskimo guide set out into the impossibly cold, twenty-four-hour night to beg help from an Eskimo village. Inevitably, a storm arose. Had he been alone, Kane would have gone the way of Cuvier's mammoths, frozen solid in a drift of snow; however, Hans Christian, his guide, had prepared for likely disaster by building an emergency igloo. When the blow came, they hid out in the ice hut. Outside, the weather was lethal; inside, it was only cold enough to make its occupants miserable. Sunrise was still a month away. Snow buried the igloo and everything else while the two explorers sat in their ice tomb knowing only that they had to stay however long the gale lasted.

After two days they dug themselves out and tried to take advantage of some brief moonlight, but the snowdrifts made travel impossible. Hans and Kane returned to their ice hole and hid from more storm, barely making it back into the igloo refuge before becoming future fossils themselves. It had been Kane's bad luck to discover a simple explanation for how fossil mammoths could have been frozen into a block of ice. Instead of being tropical animals caught in a miraculous refrigeration, they might have been cold-climate creatures that met with a routine worsening of the already foul weather.

Of course, it was more than ignorance that had set Kane starving in an igloo. There had been ambition as well. Kane was determined to make his mark, to do a splendid something before he died, and death's shadow was never far from him. During his college days rheumatic fever had damaged his heart. He survived, but his doctors warned him he could fall dead at any moment, as suddenly as though felled by a musket shot.

At first the illness and prognoses had paralyzed him with melancholy, but his father, a leading Philadelphia intellectual and jurist, urged him that if his life were to be short, he should still make the best of it and "die in harness." The words kindled something in the young man. The flames grew a little stronger when one of his teachers, William Barton Rogers, a geologist, made natural history sound interesting enough to excite Kane. And Kane then took fire with a lasting blaze when, in 1842, Alfred, Lord Tennyson published his poem "Ulysses." It was seventy lines of blank verse that he wrote shortly after the death of a close friend, his sister's fiancé. The poem presented a question everybody must answer—*Now that I understand I must die, how shall I live my life?*—and its ending struck Kane as being directed straight at him and his weakened heart:

> *Death closes all; but something ere the end,*
> *Some work of noble note, may yet be done,*
> *Not unbecoming men that strove with Gods.*
> . . .
> *'Tis not too late to seek a newer world.*
> *...for my purpose holds*
> *To sail beyond the sunset, and the baths*
> *Of all the western stars until I die.*
> . . .
> *One equal temper of heroic hearts,*
> *Made weak by time and fate, but strong in will*
> *To strive, to seek, to find, and not to yield.*

Kane's ambition to seize the day, every day, eventually led him into that snowstorm in northern Greenland. First, however, because he was a dutiful son, he earned the medical degree his parents wanted him to have. He then turned to the life of an apprentice Ulysses and set sail toward China as an assistant surgeon in the United States Navy. That was in spring 1843. In the summer of 1845, he returned to Philadelphia, having been around the world. From China he had gone on to Egypt, where he joined in some military adventures up the Nile. Then he crossed overland up Italy, over Switzerland's Alps, and into France before catching a ship for America. Back home after many adventures, his ambition burned still brighter. He had acquired some—what shall we call it?—not glory, not reputation, but a counterfeit version of them. He had become a local celebrity, and the taste of it kindled in him the ambition for more.

Celebrities seem "postmodern," but, according to the *Oxford English Dictionary*, the word, with its contemporary meaning, first appeared in the 1840s. Kane was among the first to discover that the mass press had made it possible to achieve fame without glory or reputation. The long age of the celebrity had begun. When the Mexican War broke out, Kane used his renown to gain an interview with President James Polk, who gave him a message to carry to General Winfield Scott in Mexico. Kane set off on adventures that, though minor to the war effort, won him still more celebrity. Nor did Kane's ambition cool when the war ended, despite his spreading name; he had not yet done any work that Tennyson would rank of noble nature, and perhaps he knew it.

A man of Kane's ambition and experience was a natural member of the exploration party that the U.S. Navy sent in 1850 to search for the lost ship and crew of John Franklin. By then his father was dismayed by the energy Kane brought to the task of dying in harness, and he complained, "Oh! this glory! When the cost is fully counted up, it is no such great speculation after all."

The son found that celebrity grew on celebrity. When he returned from the Arctic, the ship's captain expressed no interest in telling the general public about the frozen north. ("He is well and even excellent in the practical duties of his profession," Kane wrote of his captain, "but ignorant of all other & *proud* of his ignorance.") Kane was quite ready to fill the vacuum, and he made a speaking tour, describing his adventures to audiences in many cities. As he spoke he urged another expedition in search of Franklin, this one via the Open Polar Sea.

During these lectures Kane met and socialized with other celebrities—most notably Margaret Fox of the "Fox sisters." She was one of a pair of teenage girls who had been promoted by the showman P. T. Barnum and the newspaper editor Horace Greeley. Between them the Fox sisters and their hucksters invented modern spiritualism. Communicating with the dead, of course, is very old, but so is Shakespeare's joke:

> *—I can call spirits from the vasty deep.*
> *—Why, so can I, or so can any man;*
> *But will they come when you do call for them?*

The Fox sisters could make the dead come. They called; the spirits responded by rapping. Years later Margaret revealed the secret. The sisters made the sounds by snapping their toes the way more average people snap their fingers. The addition of curtains, dark lighting, and ample newspaper coverage raised the feat beyond the level of a local stunt by a pair of country girls.

By most standards Margaret Fox was an impossible romance for Kane, as absurd a union as Cinderella and the prince. Kane was from the well-to-do, respected, urban world—educated, traveled, a man of science. Margaret Fox was an unschooled antiscientist. Kane, however, was strongly attracted to her, and she seems to have fallen in love with him. Kane never doubted that her seances were fakes, and he was forever urging her to abandon her shows. He wrote her in what he thought was a love letter, "Am I not injuring my dignity by thus throwing away upon a person in a walk of life different from my own, feelings which she can never understand and of which she is not worthy?" But the heart's reasons overcame all pride. Kane kept chasing, and Margaret did not throw the letter in his face. The whole time that Kane prepared for his return to the Arctic he was conducting a secret romance.

If we see the couple as Kane did, with Kane as reason's pilgrim and Fox as a barker for superstition, they do seem an especially star-crossed Montague and Capulet. But viewed as a pair standing among the first celebrities created by America's mass press, their romance seems less surprising. Celebrities woo other celebrities. In the public's eye celebrities have always seemed to be interchangeable symbols of human potential. From that perspective, ignorant, fraudulent Margaret was the one who better understood the truth of themselves. She knew that they both were contributing more to America's emotional life than to its rational discourse. Margaret knew she was snapping her toes and that spirits were not literally rapping at the air. She knew too that her audience was not responding to something inherent in her sounds, but to some desire in its heart. Kane, meanwhile, did not understand so well that when he proclaimed a quest for the Open Polar Sea and when audiences of educated professionals cheered his words, they too were responding to a heartfelt dream that the world is not really the brutal, death-ridden place it seems to be.

On May 30, 1853, when Kane sailed off as captain of a ship bound for the open waters around the North Pole, the prayers and good wishes of the nation went with him as their hearts' champion. What he found, of course, was a ferocity of weather cruel enough to freeze a mammoth. Dreams in, nightmares out.

THE UPPER RHÔNE VALLEY,
SWITZERLAND, SEPTEMBER 19, 1832...

Lyell's contribution to this history is that of a politician who creates a constituency for something new. Geology in 1832 was not a well-established science. Even its name, "geology," was only a few decades old, and for many in the more established sciences of the day (e.g., astronomy, physics, and chemistry) geology still had many of the trappings of what today is called "junk science," in which facts and guesses mingle freely. Lyell worked to clear out the junk so that scientists in every field could respect the results of geology. Lyell's work was controversial among other geologists, but it appealed powerfully to young scientists.

James Forbes, for example, was a physicist who shared exactly Lyell's prejudices that the world should be explained by the actual processes known to be afoot on the globe. Twenty-one years before Kane set off for Smith Sound, Forbes was in Switzerland following the Rhône River downstream. This close to its source, the Rhône was more creek than river. Great mountains capped with glaciers lined both sides of the valley. The Alps had been discovered by growing numbers of tourists. As early as 1818, Lyell had reported that the region around Mont Blanc had received 1,400 tourists in a single year, 1,000 of whom were British. Fourteen years later the number surely had surpassed 2,000. But only the bolder travelers went as far upriver as Forbes had gotten. This part of Switzerland was still one of the most remote places in western Europe.

A Swiss named Ignace Venetz happened along and fell into a conversation with Forbes. Venetz was a civil engineer for the canton (the Swiss equivalent of a state) of the Valais. He organized the road and bridge repairs in the Alpine district, where avalanches and rock slides did constant damage. He also had developed techniques for cutting ice off glaciers to prevent the damming of streams. Venetz may have had the most detailed and practical knowledge of the central Alps of anybody then living. He was the sort of person an eager and intelligent tourist prays to meet. He could point out peaks, glaciers, and places of note on the valley floor.

As they spoke Venetz told Forbes of a paper he had written three years earlier for the *Société Helvétique des Sciences Naturelles* (Swiss Society of Natural Sciences). As a civil engineer, Venetz was stepping a bit outside his field when he argued about natural history, but he was one of those hard-working optimists who believed that if he taught himself the facts of a matter, he could make his contribution. He taught himself about the moraines that glaciers push up and concluded that the Alpine glaciers, during some former epoch now lost to the nights of time, had been much larger, producing ridges on the valley floor that still hampered road building. Forbes was intrigued, but in his journal he dismissed the idea of explaining geologic curiosities "by the Moraine of a huge Glacier" because it "seems a rather bold speculation."

Forbes's reaction was that of a scientist used to explaining all phenomena in terms of other recognized phenomena. Newton's ability to explain planetary motion by the same laws that explain falling bodies on earth had set the tone, and later physicists just naturally looked to the known for explanations of the unknown. Most geologists, however, were used to strange explanations, and they appealed to tidal waves having many magnitudes greater than the largest recorded tsunami, or they imagined global floods or continent-sized earthquakes. These ideas generally persuaded no one other than the theory's originator, and perhaps his mother. To Forbes's ears, Venetz's glacier fell into this same fantastic school. For a scientist like Forbes, Lyell's grounding of his *Principles of Geology* in actual experience held an automatic appeal, and in return, Forbes was exactly the kind of trained scientist Lyell wanted to win over.

Lyell had begun his push to create a truly scientific geology—one that, in Lyell's language, was properly "philosophical"—during the 1820s. His father had wanted him to be a lawyer, and because he was a dutiful son, he had gotten a law degree, but his own interest was in geology. He had taken William Buckland's course in geology at Oxford and had made a singularly favorable impression on his teacher. In 1824, at Buckland's invitation, the two men had toured Scotland's Great Glen together, examining all the famous lochs and the strange straight lines that scar some of the Highland slopes.

The next year, 1825, Lyell began writing articles for the Geological Society of London in which he compared recent lake deposits in Scotland with older geologic formations around Paris. This technique of comparing present-day actions in Scotland with ancient formations around Paris was already a well-established approach in geology. Buckland wrote many papers that way and taught Lyell to do the same, but Lyell became increasingly ardent that this method was the only philosophically correct method of doing geology. Cuvier, at the same time, was writing about the importance of catastrophes in natural history, a proposition that Lyell saw as completely "unphilosophical."

In his *Principles,* Lyell managed to imply that religious theorizers who traced all oddities to Noah's flood and scientific investigators like Cuvier who looked to catastrophes were all part of the same superstitious lot. Lyell had taken on a life's work, just as surely, and perhaps just as consciously, as Kane would take up his plan to be a famous hero. Lyell would be an advocate for philosophic geology, showing how all changes could be explained by the combination of time and present-day geological processes. It is likely that Forbes's meeting with Venetz came before Lyell had ever heard a whisper of Venetz's theory, but from Lyell's point of view Forbes was absolutely right in finding the idea of a great glacier entertaining and then in rejecting it out of hand.

NEUCHÂTEL, SWITZERLAND, NOVEMBER 1832...

Shortly after Forbes's encounter with Venetz, Cuvier's star pupil, Louis Agassiz, arrived in Neuchâtel, a Swiss town of 6,000 souls. Nothing better illustrates the thinness of the material threads supporting natural science in the 1830s than the wretched appointment made available to Europe's most promising student. Cuvier's sudden death had ended Agassiz's Paris career, despite the fact the Cuvier had put him in charge of his (Cuvier's) plan to produce a comprehensive study of fossil fish. There were very few jobs anywhere in the world for a naturalist, and it took all of Agassiz's connections to land even this post of so little promise. His mother had reported that Neuchâtel was about to establish a *Collège*—what Americans would call a high school—and planned to open a museum of natural history as well. Anybody who has ever visited a small-town museum can imagine how thoroughly mediocre an acceptable museum curator could have been for Neuchâtel, but Agassiz decided to try for a position as a teacher at the new school and as the organizer of the new museum.

He spoke to Humboldt, and Humboldt then wrote to Prussia's minister of education, who was also his brother. In those years Neuchâtel was the last Swiss canton that did not enjoy its own liberty. Neuchâtel was still a Prussian principality, although the canton lies along the French-Swiss border and is nowhere near Germany. Humboldt also wrote to the Prussian governor of Neuchâtel and to his friend Leopold von Buch, professor of geology at the University of Berlin.

With this energetic backing, Agassiz won the posting, but it was left to him to turn the position into something more than a trivial bit of patronage. Agassiz, then twenty-five, arrived with so much determination and ambition that he might have conquered London or Berlin. Neuchâtel was at his feet at once. He knew the local country well. He had been born (May 28, 1807) in the village of Môtier, only about a dozen miles from Neuchâtel. This part of Switzerland lies west of the Alps and east of another, smaller chain of mountains called the Jura. Besides the distant mountains, the dominant feature in this area is Lake Neuchâtel, a long, narrow water with a slight curve along its shoreline. With the Alps so near, this valley has no special reputation for beauty, but it is an elegant and charming natural setting.

Agassiz's father was a Protestant minister who was himself the son of a minister, the grandson of a minister, and in fact the descendant of all ministers going back to his great-great-great-grandfather. Surely there must have been some expectation and hope that Louis would become the seventh in the proud line, but his interests soon pointed elsewhere. Especially, he was curious about fish and nature, and his parents provided the education necessary to further such devotion. But there were no livings to be had as a natural scientist, so as a dutiful son, Louis obtained a medical degree, although he never practiced medicine. Instead, he used his degree as an excuse to go to Paris, claiming he was going to study the cholera outbreak, but really seizing an opportunity to impress himself upon Cuvier, Humboldt, and the scientific world.

Now he was back in Neuchâtel, where he might have been a doctor, as a teacher. Many years later, when he was a father with grown children of his own, his children referred to him as the "steam engine," a go-go machine that never tired. He was that way the moment he reached town. Betraying no disappointment or doubts, he leapt into action like a locomotive with no brakes and infinite fuel.

He began teaching in November. By December he had also organized the Neuchâtel Society of Natural Sciences, though it is hard without laughing to imagine what the society was like with its six amateur members and genius leader. Yet over time the group became valuable. Its members, at least some of them, were hardworking and self-confident. Two years later, in the summer of 1835, Lyell passed through Neuchâtel and found that, although Agassiz was absent, several local people could give him useful guidance to

Neuchâtel's geology. Agassiz had also gone immediately to work on his investigation of fossil fish. He even had an illustrator in his employ, an artist named Joseph Dinkel, whom Agassiz had first hired when he (Agassiz) was a student in Munich. It seems fantastic that a student could hire a personal artist out of an allowance provided by his none-too-rich parents, but Agassiz put his work far ahead of his income. Many years later Ralph Waldo Emerson put an approving anecdote in one of his notebooks: "told of Agassiz that when some one applied to him to read lectures, or some other paying employment, he answered, 'I can't waste time in earning money.'"

Indeed, the newcomer to Neuchâtel was fully engaged in nonpaying pursuits. Fossil fish, for example, earned him nothing, yet they occupied his mind even while he slept. One of his projects required that he remove a fish from the stone where it was embedded. The fish was partially exposed, but the rest was hidden. Agassiz's job was to chip away the stone without destroying the fossil, a task possible only if you know where the fossil should be. But what does a new fish look like? Cuvier had developed a means of predicting anatomy from the available bones, but it took a genius of Cuvier's level to apply the system. Agassiz struggled with the stone for days. One night he dreamt of the fish's shape, but the next day, when he faced the stone, he could not remember what he had seen. He saw the fish again the next night, and again he forgot its appearance during his waking hours. The third night he took paper and pen to bed with him, and, when he again dreamt of the fish, he woke and at once sketched it out. The next day he chipped the fossil from the stone and found that it matched his sketch precisely.

During this early period in Neuchâtel, Agassiz seems also to have clarified his religious opinions. As the heir to a line of ministers going back almost 200 years, he was deeply aware of the contradiction between natural science and biblical history. The chronology of Genesis was false. Death had come into this world long before Adam's Fall. There had been no global deluge. These doctrines were shocking to many people in the 1830s, and they still have the power to astonish some naïfs even today, but Agassiz did find some doctrinal reassurances in his fossil fish.

First, they reaffirmed the tradition of a Creator. Agassiz joined Cuvier in rejecting Lamarck's evolutionary ideas, finding "that species do not pass insensibly one into another, but they appear unexpectedly, without direct relations with their precursors." Second, the fossils were so well fitted to their needs that they were surely the result of intelligent planing. The Creator had some purpose in making things as He did, and, as a result, our lives have meaning.

Living, as we do, after Darwin, we must conclude that although Agassiz was an intense young man, ambitious and with a head full of facts, he was wrong about natural history's big picture. But being wrong, even in science, turns out to be a not-so-terrible way to begin. All three key people in this history of the Ice Age's discovery began by being wrong about great ice. Agassiz had been born and raised in Switzerland, and he was working only a short journey from the largest Alpine glaciers, yet he seems never to have noticed the ice. Elisha Kane was heading out to the top of Greenland, expecting to find that the ice naturally thinned out and disappeared as he moved closer to the North Pole. Charles Lyell had visited Switzerland and hiked across the

Mont Blanc glacier, but somehow he missed the fact that ice rearranges the landscape. Yet these three men would become the ones who made the Ice Age comprehensible to the whole world.

If they had been machines, their task would have been hopelessly wrong footed, but each of them also had a determination to do something right by the world. They were proud and ambitious for themselves, but they were not merely young opportunists. It is probably not a coincidence that each of them had deferred to a parent's wish and gotten a professional degree. They each brought to the efforts ahead a sense of obligation to something more than themselves.

Part II

Trading Ignorance for Action

ST. JOHN'S, NEWFOUNDLAND, JUNE 17, 1853...
Elisha Kent Kane is the figure whose sense of larger obligation is the most doubtful. He was the one most focused on his own name and glory. Officially, he had a specific purpose—find Sir John Franklin's party, dead or alive—but that goal was always more MacGuffin than real. It was an excuse that adventurers used to get themselves north and put their names onto maps; however, it was so romantically grand an excuse that it blinded Kane to the more ordinary motives of his party members. They needed work. They were curious. They thought they could help out. Kane had set sail without one of the primary requirements of a leader: an instinct for why his people follow.

Kane's ship, the *Advance,* had stopped in Newfoundland to load more supplies. The Newfoundland governor supplied the party with a team of dogs to help pull the sleds. Anyone experienced with even a solitary full-sized dog knows that you have to be direct and forceful if you are to stay its master. These Newfoundland dogs were wilder than any domestic breed, and they were far from solitary. It was impossible for any novice to be firm enough with this pack of half-wolves, and the arrival of the dogs marked the departure of peace.

St. John's provided one last chance to send mail home. In one of Kane's letters he joked with his sister about the pleasures of leadership:

> *Soon—disgustingly soon for our better natures—we learn to take to ourselves the perquisites of accident—and this trade of authority sits so easily upon me that I feel as if to be toadied was a natural and inevitable province of my peculiar self. The best piece at table—the best places on deck—the smile at a bad joke—the affected comprehension of a good one—the deference to an absurd idea—the jump at a request or suggestion—all these...take as naturally as the filling of a Burgomaster's pipe—smoking, puffing, pausing and casting away, just as it suits or does not suit I myself.*

This ironic view of leadership can serve among old pals who have agreed that someone must be in charge of organizing the fraternity ball, but it works poorly on board a naval vessel. The crew cannot be let in on the joke, although it soon catches on that something is out of tune. Kane's crew noticed quickly that its captain was slow to assert himself when assertion was expected. Contributing to Kane's poor impression was his physical weakness. He did not have the presence of a hero. One of the crew, Henry Goodfellow, wrote that at first Kane "had the subdued look of a broken-down invalid." By the time the ship sailed, Goodfellow added, "he had recovered in a great degree the tone of his bearing, but he was far from either well or vigorous."

After two days anchored in Canada, the brig continued on toward Greenland and the Arctic Circle. The Western

Hemisphere above St. John's is, in most imaginations, compressed and hazy. Less than 1 percent of New Worlders live above that point, but St. John's, on the forty-seventh parallel, is almost as close to the equator as it is to the North Pole, and Kane was bound for a point above magnetic north. The ship needed almost two weeks to sail straight north, to the village of Fiskernaes at Greenland's bottom.

Fiskernaes had the look of a frontier settlement with a few shacks scattered in where-you-will fashion. In the background rose splendid mountains that offered the newcomers their first glimmer of the terrible Greenland ice sheet. The unexpected arrival of a ship brought the whole village, Danish pioneers mostly, out to gawk. And as was also true of the U.S. western frontier, the explorers had moved beyond the civilized map-reader's imagination. Just as eastern Americans had no clear notion of the roadblocks posed by the western frontier's endless lines of mountain behind mountain behind mountain, southern map readers could not guess how travel through this northern frontier is impeded by a strange blockade, what Kane called Baffin Bay's "essential feature," its middle ice.

If you look at a map, the easiest course north through Baffin Bay looks to be straight up the middle, but that way is as impassable as the Himalayas. Ice floes and bergs have come down from the north and settled into a kind of permanent thicket. Ships cannot cross wherever they choose between the Canadian and Greenland sides. Whalers had identified a couple of passages through this middle ice, and knowledge of these routes was as crucial as knowledge of the passes through the Rockies. For the most part, ships in these waters did what Kane did: follow the Greenland coast on up.

As the party made its way along the northern frontier, it obtained more dogs, these even wilder than the Newfoundland ones, and it added an important man to the crew. Kane hired an Eskimo, named Hans Christian, to sail with his ship. As proof of his skill, Hans speared a bird on the wing. Impressed, Kane took him on at once. Hans's knowledge of hunting and cold-weather survival would prove so crucial to the expedition that it is hard to conceive of anybody's survival without him. Yet he joined the party at the final minute and in an offhand way. Kane thought it might be useful to have a hunter provide food for the dogs. There was no suggestion he might prove valuable to the two-legged members as well. Perhaps no detail of the expedition makes the continued ignorance of great ice more clear than the happenstance manner in which Hans joined the crew. The party had not guessed it was going some place so difficult that its boat might be entombed in ice.

Kane sketched scenes as the ship cruised along the coast, and the details show he was alert to the geological effects of ice. He drew a section of coast and wrote that the cliff side between two points, which he marked, was a perfectly plane surface reflecting the abrasion of ice.

The last trading settlement on the coast, about two-thirds of the way up Greenland, was Upernavik. A couple of fishing villages lay a bit farther north, and then the frontier disappeared. Settlers had not gone farther up the coast and were not going to go farther up. Even most whalers did not go much beyond Upernavik. The middle ice opened here, providing a route to the bay's Canadian side. Meanwhile, on the Greenland side, not far above Upernavik, lies Melville Bay. Greenland's interior ice sheet presses right to the shoreline here, making further settlement impossible and also filling the water with so many icebergs that ships routinely turn away. This is the area launching all those icebergs that Lyell had mistakenly placed on Greenland's eastern side. It is from here, most likely, that sixty years later, the *Titanic*'s fatal iceberg broke off. Kane's route passed directly through this bay.

The *Advance* stopped at Upernavik for a final glimpse of civilization and took on one last crewman, an English-speaking Dane named Carl Petersen, who knew the Eskimo language and even knew how to drive a team of sled dogs. Petersen grumbled in his journal that the group appeared to be poorly led, but he came, and then the brig sailed on toward Melville Bay.

LONDON, FEBRUARY 24, 1834...
Charles Lyell began his acquaintance with Louis Agassiz by writing him a letter in French:

> My dear sir,
> It is with the greatest pleasure that I send you good news. The Geological Society of London has commanded me to inform you that this year you have been accorded the prize legacied by Doctor Castor.... Your work on fish was considered by the council and the officers of the Geological Society as worthy of the prize.

Thirty guineas were awarded for Agassiz's first volume (of, eventually, five volumes of text and another five volumes of plates) describing fossil fishes, a book that impressed the best naturalists in Germany, France, and Britain. The British prize-givers had been especially impressed by the work's detail, the depth of its natural history, and the beauty and precision of its illustrations. The extravagance of hiring a personal illustrator now proved its worth. With Agassiz's book, readers could see exactly what these fossil fish looked like and learn how they fit into the earth's story. There was perhaps only one publisher in Europe willing to spend the money and attention needed to get this book right, and that was Agassiz's own house, founded in Neuchâtel to print whatever he chose in as elaborate an edition as he could design.

Lyell's letter of appreciation began a lifelong entanglement between these two men. It became one part friendship, two parts rivalry, a heaping cup of shared respect, and a dash of bitters. The bitters especially produced an enduring aftertaste; however, the initial ingredient set forth in this letter was all respect.

Lyell, the sworn enemy of Cuvier's catastrophism, and Agassiz, Cuvier's devoted heir, did not seem like natural colleagues and back-scratchers, but Agassiz's work with fish did more than fill in an empty spot on the map of paleontological knowledge. The Geological Society, for which Lyell served as secretary to foreign lands, did not award Agassiz a year's wages for an unskilled worker to encourage ichthyology. Its members saw his work as providing major geologic data. From the British perspective, nobody was doing more important geologic work than Agassiz. His research provided a key to reckoning the relative chronology of geologic events.

Fossil research was the hottest breakthrough in science. Like modern DNA research, it promised to reveal a history that had seemed beyond all knowing. By the late 1700s, geologists understood that the earth's layered structure provided the clearest indication of which features are older than others. Almost any cutaway view of the earth—from a riverbank to the Grand Canyon—shows strata, layer above layer above layer. And common sense insists that, say, a seam of coal is older than the layer of slate that lies above it. Presumably earlier thinkers too could have figured out this much if they had not thought the whole world had been created at one stroke.

There is a coincidence so unlikely that I hesitate to mention it, lest readers suspect they are holding a novel, à la Dickens, rather than a history. The man who argued effectively for the importance of the earth's strata, a Russian named Mikhail Lomonosov, was also the first person to try to reach the North Pole and was a firm believer in the Open Polar Sea. Thus, he was the grandfather of both the geological and Arctic strands in this history.

Geologists of the early 1800s were kept busy mapping out the earth's strata, and, for the first time, they were able to give a relative chronology of the world. A man like William Buckland recruited promising minds into geology by taking them to a hill overlooking the Thames valley and pointing out all the details in the riverbank and terrain that told the story of the scene's history. The bright student would stand amazed; all these lines and layers that had previously passed unnoticed were telling of a lost world. It was like seeing forgotten ancestors in the face of a child.

But strata by themselves are ambiguous. England has limestone, and the mountains along the French-Swiss border are largely composed of limestone. Do these separate limestone beds trace to the same era? In 1815, less than twenty years before Lyell first wrote to Agassiz, a British geologist named William Smith argued that fossils provided the key to linking strata. The strata show the relative chronology of a region, and fossils in the strata show the order in which species appeared and became extinct. If you wanted to compare the dates of English and Swiss limestone, the secret was in comparing their fossils.

Geology's first spectacular success in relative dating lay in overturning various "proofs" of Noah's flood. English geologists, notably William Buckland, had found ample evidence of what they called the "diluvium," remains of Noah's flood. This consisted mostly of stray boulders and odd scarring of the earth that seemed to be the debris of a huge flood. By 1830, however, it had become clear that the diluvium did not trace to a single era and could not be the result of a single, stressful, forty-day event. If a science is not a science until it has overthrown some famous biblical passage, geology had come of age.

Lyell's letter to Agassiz was England's acknowledgment that the Swiss teacher was in the forefront of the geological revolution then under way. His work mattered because so many strata combined fish and mollusk remains. Agassiz's book cataloged many of these fossil fish, listing which had come before and which after others. Agassiz was also able to say which fish were freshwater and which were marine species, enabling geologists to determine whether they were looking at the remains of a lake or a sea.

Britain's geologists were startled to find a twenty-seven-year-old scholar who already knew so much natural history that he could provide a detailed key to water-deposited strata. England at that time was full of amateur geologists who were working out in detail the strata of their home counties, and suddenly, thanks to Agassiz, they had clear illustrations and facts to help them resolve the chronology of their local formations. For providing so valuable a guide, even Lyell was willing to honor a disciple of Cuvier.

The prize that the Geological Society awarded Agassiz was one of many that the fossil fish work brought him. He was honored by groups in Berlin and Paris as well. But British geology was entering an especially rich period, as its amateur members worked out local chronologies and its leading professionals like Buckland, Lyell, and Roderick Murchison glued the chips together into a full history of their island's geological and biological history. Agassiz, over in Switzerland, was welcomed into British geology's golden day during its morning hours.

LUCERNE, SWITZERLAND, JULY 1834...
The Swiss Society of Natural Sciences called its annual meeting to order, and one by one, members began presenting their learned papers. Agassiz was there, of course. By then he had been in Neuchâtel for almost two years and had established himself as the human steam engine who was always conceiving new projects, always spending money he did not have, always finding new ideas to chase down. His Neuchâtel Society of Natural Sciences provided him with a dignified venue for presenting his papers, and he took full advantage of the situation. His publishing house was busy too. Humboldt had generously persuaded King Frederick William III of Prussia to support Agassiz. The king is remembered in history, on his best days, as a mediocrity, but Agassiz had little reason to complain about his monarch.

Somehow during this time Agassiz also found a spare minute in which to get married. His wife, Cécile, was the sister of his schoolmate and best friend, Alexander Braun. She quickly discovered the price of marriage to a great man: His thoughts do not linger over the domestic. Agassiz was always busy and often on the road. He was perfectly capable of telling his new bride that he was off to Lucerne for a conference and would be back when he got back.

The other scholars at the Lucerne meeting were not so driven, but they too were making sacrifices for science. Europe had very few men who could make a living as a scientist, especially one interested in natural history. Full-time naturalists tended to be wealthy eccentrics. For example, Germany's most important geologist was the independently wealthy Leopold von Buch. Besides intelligence and energy, von Buch used his personal fortune to finance expeditions to distant sites. Without such a purse, a naturalist needed a job and enough love for science to do what he could on the side.

At the Lucerne meeting Jean de Charpentier was a typical underfinanced scholar. By day he was the director of the salt mine in Bex, not far from Switzerland's greatest peaks. In his free time he was a geologist with connections all over Europe. He had gone to mining school (in those days the only place to learn geology and mineralogy) with both von Buch and Humboldt. He was a member of scientific societies in Breslau, Dresden, Hanau, Lausanne, Leipzig, and Marburg. He was also a foreign member of the Geological Society of London and many natural history societies in Paris and other parts of France. So when Charpentier presented his paper, he was not reading the ruminations of a Sunday naturalist. He was a respected, recognized student of natural history.

Yet his paper was not received respectfully, for in it he declared himself a convert to Ignace Venetz's theory that Switzerland had once been buried under a glacier. He presented his evidence clearly. Indeed, Charpentier's paper was the first document to assemble the classic evidence of ancient glaciers. He recited his facts well, but even a fact man like Agassiz paid no attention. Agassiz ignored the paper even though he had known and respected Charpentier for years. Before Paris, before his higher education in Germany, Agassiz had gone to secondary school in Lausanne, on Lake Geneva, and Charpentier had been one of his teachers.

Today's reader wants an explanation for the contempt that greeted Charpentier's paper. We are puzzled because the idea of an ice sheet seems eminently plausible to us. But imagine the reaction today if something inherently absurd were argued logically and factually—perhaps somebody argues that back in Roman times cats barked and dogs meowed. It really would not matter how many technical facts about barking and meowing the scientist presented or how much evidence there was that Cicero's dogs could not have barked; most people would say there has to be some other explanation for these facts. We would not spare the time to inquire further into the eccentric's own assemblage of details.

The antiglacier, anticold attitudes of that period were not just mere garbage in. They were garbage packed in and tamped down. These were prejudices so inherent and so undisputed that the brightest scholars of the 1830s responded to the glacier idea the way we would to the dogs-once-meowed thesis. The marvel of July 1834 was not the banality of the audience, but Charpentier's own ability to overcome those prejudices in himself. He appeared before those naturalists like a reformed sinner who has seen the error of his ways. Charpentier had first heard the glacier idea almost twenty years earlier, in 1815. He had met a mountaineer named Jean-Pierre Perraudin, who believed that the Rhône valley, which runs through the heart of the Alps, had been filled with a glacier that extended as far as the town of Martigny. (Perraudin had never been beyond Martigny and could make no claims about that larger world.) His argument was that there were too many stray boulders scattered about the valley floor. Charpentier recalled, "Although I agreed with him on the impossibility of transporting erratic boulders by water, I nonetheless found his hypothesis so extraordinary and even so extravagant that I considered it not worth examining or even considering."

During the intervening years, between Charpentier's talk with Perraudin and his presentation at Lucerne, Charles Lyell had patched up the difficulty of spreading stray boulders by water alone. Lyell had learned that icebergs sometimes carry boulders along for the ride. When they melt, they drop their stone passengers at some random location. During past inundations, when today's dry land was underwater, icebergs could have carried these boulders and dropped them. Today these boulders would be open to view. To our ears the explanation seems far-fetched, but in the 1830s it sounded more plausible than the theory of fantastic glaciers. Agassiz thought Lyell was probably correct.

Charpentier had changed his own mind because Venetz had personally taken Charpentier to sites in the Rhône valley and shown him what he himself had seen. *Look at this, look at that,* Venetz had pointed, and he insisted he was pointing out the remains of a glacier. Charpentier had recognized Venetz's evidence. Forced to choose between his eyes and his prejudices, Charpentier had gone with his eyes.

That act of recognition was how Charpentier had gotten from garbage in to knowledge out. When Agassiz heard the Lucerne paper, however, he experienced no such recognition. Charpentier would have needed a poet's art to inspire that deep a response. Instead, he was a scientist organizing facts, and no one took him seriously.

OFF WILCOX POINT, GREENLAND, JULY 27, 1853...
An ice fog hid the waters as the *Advance* arrived at Melville Bay. Kane, staring intently into the white darkness, could see little of the floating ice that surely surrounded him. At any time an ice mountain might pop out from the frozen mist. On his earlier voyage he had seen right here a 700-foot-high glacier calving the stuff that gave this water its nickname, "Bergy Hole."

The question—impossible to answer in the fog—was whether the bay was navigable that season. Whalers generally tried to hug the shore, making their way through a narrow open lane while deftly avoiding any bergs that fell from the glaciers, but that technique only served if the sailors could see and if the shore lane was still open.

Too late, when the fog lifted, Kane saw that the coastal ice had melted and broken up already, turning the shore route unpredictably treacherous. Even if they managed to pick their way through 10 or 30 miles of coastal route, they could not be sure of what lay ahead. This route traced 150 miles of glacial faces. Nor could they rely on the escape route remaining open behind them. Isaac Hayes, the ship's surgeon, described the scene as "an immense, impenetrable wilderness, which grows worse and worse." Kane gave the order, and his brig turned westward, away from the shore.

The ice packs in this deeper water formed a confused mix of floe and berg. The floes had drifted down from the north and from whatever lay in that unknown land. The bergs—most of them—had come from the bay's eastern shore. To penetrate and cross the mix, Kane tried to take advantage of one of the jam's oddities. The ice floes were drifting south while the icebergs immediately next to them sailed northward. The surface current moved toward the Atlantic and took the ice with it. But the icebergs—many rising 100 or more feet above the waves—reached down hundreds of feet below the surface. Down there a deep current ran to the north, pushing the icebergs along with it. Kane's plan was to fasten the *Advance* to an iceberg and take a tow northward.

The scheme was not danger-free. First, tying oneself to an iceberg takes risky work. The crew once spent eight hours struggling to fasten the ship to ice. Second, you cannot trust icebergs in their conduct. Soon after the men congratulated themselves on securing an iceberg, they were startled by loud sounds and the sight of walnut-sized chips of ice flying off the berg. The thing was generating its own hailstorm. The crew members scrambled to detach themselves from whatever was going on and just barely managed to get away when the whole iceberg came apart as though struck by artillery.

The bay held other surprises. Far from shore the crewmen saw two polar bears riding the ice, and they shot both of them to feed the dogs. This was the season of the midnight sun, and it never turned completely dark. The sun circled and dipped toward the horizon, but then it came back up. As the sun slipped toward its faux set, it reddened the sky around it while turning the rest of the world gray. The light reminded Kane more of an eclipse than of twilight.

In the midst of this secret world, the *Advance* struck an iceberg and nearly foundered. One of the expedition's boats was destroyed, the figurehead was split open at the breast, and the jibboom was swept away. The hull was not cracked, however, and the crew rose to the occasion by exhausting itself and saving the ship.

Yet for all their dangers, the drifting icebergs seemed the best escorts to Greenland's unmapped north. Kane hooked on to one and then another. His sextant proved that they were slowly—sometimes barely at a mile an hour—moving in the right direction. He was alarmed to see, however, that his berg had pulled them closer to shore, and they found themselves facing "a blank wall of glacier."

Whalers had known of this glacial wall for centuries, but most scientists and naturalists remained ignorant of it. Americans and Europeans, if they thought anything at all of glaciers, pictured those of the Alps. Kane knew the Swiss glaciers too, having crossed that way during his return home from China, and he knew that comparing Alpine glaciers to those along Melville Bay was like comparing peas to watermelons. In describing one of the Greenland glaciers he studied during his first Arctic expedition, Kane had said that Europe's most famous glacier—France's Mer de Glace on Mont Blanc—"might repose on the slope of a single ice hill," whereas the Aletsch glacier, Europe's largest, could fit "in one of its ravines." The whole coast here was packed with fjords through which oozed Greenland's ice sheet. That interior ice made these glaciers unlike anything known to Alpine experts. The Mer de Glace and the Aletsch were separate bodies, just as the Hudson and Delaware Rivers are separate waters with their own distinct sources. But the glaciers of Melville Bay come from the behemoth ice block that buries the Greenland interior. These glaciers were not separate phenomena any more than the many streams and mouths of the Mississippi Delta are distinct waterways. Naturalists, like those who heard Charpentier in Lucerne, who tried to imagine great ice and its effects without picturing something like Melville Bay's blank wall of glacier were lost. Their knowledge of Switzerland blinded rather than enlightened them.

Eventually the *Advance*'s free iceberg tow began to fail. The floes built up as they moved north, and finally their downward push overwhelmed the iceberg's upward pull. The iceberg began moving south with the rest of the waters, so Kane ordered the *Advance* to continue the rest of the way through the ice maze under its own effort. Kane trusted none of his crew to find a passage through the ice, and he spent the next days scrambling between the crow's nest and the wheelhouse, pointing the route from one opening (called a "lead") to another. Yet Kane aspired to poetry and took the time to note the effect of the midnight sun as it made "the ice around us one great resplendency of gem-work, blazing carbuncles, and rubies and molten gold."

On August 4, they passed beyond Melville Bay and sailed back into reasonably open water. It had taken just over a week to cross this barrier that almost nobody ever crossed. Whalers seldom came all the way to the top of the bay. The

Franklin expedition had probably never gotten this far north. (It had last been seen back at almost the precise location where the *Advance* had smacked the iceberg.) Kane's first Arctic exploration had turned west before reaching so high a latitude. And Kane was a ship's doctor, not really a seaman, yet he had understood the ice well enough to sail beyond its barricade. The rest of the sailing would never be so dangerous; although, of course, at some time he was going to have to get back south across that same Bergy Hole.

LONDON, FEBRUARY 19, 1836...

Charles Lyell rose as president of the Geological Society and addressed the group's anniversary meeting. It was the society's annual showcase event, drawing out-of-towners and many distinguished Londoners who skipped the regular meetings. Cornish and Yorkshire merchants who enjoyed monthly excursions in the countryside to look for fossils had come to the great city for this special meeting. European geologists too had come for the occasion.

Lyell reckoned he had devoted a total of fifty days to preparing this speech, yet nothing especially memorable escaped his lips. He did, as was customary, provide a detailed summary of the previous year's accomplishments and ideas. In the course of this survey he mentioned Charpentier's theory of a great glacier that had covered the Alps. Lyell said that Charpentier "informs us that [his theory] is merely a development of one first advanced by M. Venetz."

Presumably one could also say that Newton's theory of gravity was a development of ideas first advanced by Johannes Kepler, who was merely expanding on a notion proposed by Copernicus. Whenever a speaker wants to downplay an idea, it is rhetorically helpful to suggest that there is nothing new in it. (If there *is* something new, characterize it as "novel.") Lyell was using a politician's ploy—stating something for the record while dismissing it at the same time. Nobody who heard Lyell's reference would be inspired to seek the paper out. Those who had never heard of Charpentier's idea would let this reference pass like an express train that does not stop. Those who had heard something now heard they need not bother learning any more. Agassiz was in Lyell's audience that evening. Probably he was the only person in the room who had heard Charpentier deliver his paper, but Agassiz agreed with Lyell on this point and did not question his dismissal.

The work that Lyell did dwell on concerned Darwin's letters from South America. Charles Darwin had been on HMS *Beagle*, sailing around the world, since the end of 1831. The ship had spent most of 1835 exploring South America's western coast from the Strait of Magellan north to Peru. With hindsight, we focus on Darwin's visit to the Galápagos Islands, where he made the observations that would shake the world; however, Lyell focused on Darwin's extensive study of the Andes Mountains. Darwin concluded that the Andes had been created over a very long time by causes still working today. The Andes, said Darwin, were still rising. The reports sounded as though they had been written by Lyell himself, and no wonder. Darwin was Lyell's most avid disciple.

In his address Lyell did, briefly, what he had done at length in his book: He summarized the whole of recent geology, stressing the work (like Darwin's) that supported his theories and downplaying ideas (like Charpentier's) that did not. Charpentier's theory was especially abhorrent to Lyell. It claimed there were geological mechanisms (glacier action) not listed in his book. It associated itself with French—that is, catastrophic—theories of mountain building, and it proposed glaciers of a size and force unlike any in the modern world. (Or so Lyell thought. He knew nothing, as yet, of the Greenland ice sheet.) Given Lyell's disdain for Charpentier's ideas, as well as the scorn that had greeted the paper when read aloud in Lucerne, it seems surprising that Lyell bothered even to mention them. After all, he was not required to notice every geology paper published.

Apparently Charpentier's ideas were generating some interest. His paper had been reprinted in Paris, in the summer of 1835, and a German translation had just been published in Zürich. The German version especially was reported to have caused a stir. Lyell's reference acknowledged the rising interest, but his summary knocked it back down.

After his speech the meeting proceeded to other business, including the handing out of awards. Agassiz was given a medal, his second in two years, for his work on fossil fish. Cheers filled the meeting as Agassiz was called forward. He was already one of Europe's most honored naturalists.

BEX, SWITZERLAND, SUMMER 1836...

In the 1830s, Jean de Charpentier had turned his home into a kind of summer salon for Europe's savants, and two years after the presentation at Lucerne, Agassiz came with Cécile for a visit. Cécile especially needed to enjoy herself. She loved her husband but not her circumstances. Trapped in Neuchâtel with its small-town bourgeoisie, Cécile thought Switzerland was stifling. Her husband's patrons included even the king, yet Louis spent little of the money on his family, which was growing. The previous December their first child, Alexander Agassiz, had been born. Louis was not wholly indifferent to his wife's needs, so, casting about for a solution, he remembered that Charpentier's wife was, like Cécile, a German, and he thought she would enjoy the company of another German. Besides, Bex was in the Rhône valley, astride an old route that predated Caesar, and stood at the entrance to the Valais, the heart of the classic Alps. It was sure to be a lovely place for a vacation.

Mountaineering was becoming a sport—many of the Alpine peaks would be first scaled in the 1840s—and the romantic movement had transformed people's perception of mountain scenes. Instead of being dark, dangerous blockades, impassable mountains now seemed majestic, beautiful, and worthy of contemplation. As of yet, however, few Alpine resorts existed, so people needed connections before they could spend a summer's romp in Europe's most famous mountains.

So Agassiz went to Bex without an agenda or even a checklist of must-see Alpine sights. Perhaps the human steam engine even believed himself capable of enjoying a summer without working too terribly seriously. Bex itself had a few one-star mysteries to pass the time. For example, there were strange potholes at the base of the escarpment. They looked like the kind of thing an ancient waterfall might have left, pounding holes into stone, but no river could ever have found its way above Bex. Alpine drainage runs straight down the mountain slopes, and rivers seem doomed to flow where the Rhône flows, right through the lowest part of the valley.

Potholes are subtle wonders. A more obvious puzzle was scattered against the mountain wall across from Bex, in the town of Monthey. Scattered amid the woods there stood the mysterious "blocks of Monthey," huge granite boulders that boasted 30, 40, even 60 feet on a side. Even without any human embellishments, the Monthey stones were as awesome as the carved heads on Easter Island. Many of the blocks balanced on top of one another as though raised by a tribe of Goliaths, since lost to history.

One of Charpentier's other guests that summer was Ignace Venetz, and both of them agreed that a glacier had brought those blocks. Agassiz was still a Lyellist in this matter and preferred an explanation of water and icebergs. A glance at Bex's geography justifies Agassiz's position. There are no glaciers in or around Bex. Yes, a bit over 10 as-the-crow-flies miles away is the glacier atop Les Diablerets Mountain, but glaciers are not crows, and the Les Diablerets glacier would have to go down and up a few times before reaching the neighboring villages of Bex and Monthey. Meanwhile, the Rhône flows right between the two places. Surely catastrophic flooding is more likely in this area than catastrophic glaciers. So Agassiz was confident that he was being the reasonable one, and in that mood of certainty, he challenged Charpentier and Venetz to a geological duel. The three would go into the Alps and see who was right.

So much for a summer's relaxation. Agassiz spent five months exploring the Swiss interior, gaining a full mastery of the facts about the glacier-versus-water dispute. He went deep into the Rhône valley, traveling as far as Zermatt up in the mountains by the Matterhorn, and he climbed the slopes of Mont Blanc to see the Mer de Glace, the very glacier Kane would later dismiss in contrast to Greenland's great ice. He also scaled Les Diablerets, to examine its glacier, and he studied many details of the floor and sides of the Rhône valley.

Cécile, by then, knew that her husband was apt to leave her and her son behind while he dashed off to explore some phenomenon of nature; but did Charpentier and Venetz realize that they had lassoed a locomotive when they took Agassiz's challenge? If Agassiz had been right about water, perhaps they could have conceded early on. But as the exploration continued, Agassiz saw that his rivals had much on their side, and he pressed on, a tornado with a purpose.

Agassiz had discovered that superficiality lay on his side, but the details showed only arguments to support his guides. At Bex, the Rhône flows north between two steep mountain walls. The peasants in the region are well experienced with flooding, and farming on the valley floor had become possible only after a system of dikes and canals brought some river control. A dozen miles south of Bex, at Martigny, the valley pivots 90 degrees and goes off to the northeast. This abrupt turn, and the valley's narrow squeeze at this point, would seem to preclude any overgrown glaciers from pressing on toward Bex.

Beyond Martigny, the valley and its extension run beneath the most famous Alpine peaks. To the north is the Bernese Oberland (the Bernese Alps) range, which includes the Jungfrau (13,642 feet) and the Finsteraarhorn (14,022 ft). On the valley's southern side are the Pennine Alps, home of the ancient passes between western Europe and Italy and of the famous Matterhorn, a finger of naked rock that points 14,691 feet into the sky.

Anyone traveling alongside the Rhône's course is likely to be impressed, even frightened, by this display, but glaciers are not what impresses. The Swiss Alps actively discourage travelers from imagining a world covered in glaciers. The Valais is a showcase of geological power, but all the power appears to be with the mountains. The Valais glaciers are swatches of white set high on individual peaks. This place has none of the look of Melville Bay, where glaciers dominate the mountains.

If anything, glaciers in the Valais are a blessing. All of Switzerland's important rivers, including the Rhône, begin as streams coming out of glaciers. Because of the high mountains on all sides of the valley, it hardly ever rains; and if it were not for the Alpine glaciers, the Valais would be as dry as the Nevada desert. So the first clue that the Swiss glaciers are more powerful than they appear lies far below, where they look like distant snowy fields. Glaciers are the source of the valley's life.

Agassiz, however, was never one to settle for distant views and philosophical meditations. At Chamonix, on Mont Blanc, he saw a glacier up close. The Mer de Glace looks less pure when you are on it or beside it than it does from miles below. For one thing, it shovels dirt in front of it and alongside of it. Geologists are used to looking at a landscape piled up in tidy layers. They are comforted by the many strata exposed by, say, the Grand Canyon walls, and they would be dumbfounded if some planetary orbiter sent back pictures of a great canyon somewhere in the solar system in which the walls were a jumbled mess. The moraines shoved up by glaciers are exactly that: an unstratified mix of dirt and stones. A moraine is the same sort of random heap of junk that you will see on a construction site, where bulldozers shovel everything into a pile. And at night, when the crew has gone home, the pile stands as evidence of ongoing work. The same holds true for moraines around a glacier. Near Chamonix, besides the front moraine at the glacier's head and the lateral moraines alongside the Mer de Glace, you can see abandoned moraines testifying to a past when the glacier reached a bit farther than today. Anyone who has ever lived around a glacier has seen that they advance and retreat.

Agassiz at Chamonix could see that glaciers are not just fields of snow. It is a mistake to speak of the Alps as "snow-capped." Snow that sits quietly and looks pretty can cover a low hill, like Scotland's Ben Nevis, but Alpine peaks have something else. The glaciers are hard ice structures. They have a dusting of snow along their tops, but these flakes, even when frozen together, have such a different nature that they are called "névé" to distinguish them from the glacier proper. Névé is snow on a glacier that has not yet completely turned into ice. Like snow, it is shiny and white, and, sitting as it does on top of the glacial ice, it is the névé, not the glacial ice, that makes the Alps glimmer.

The difference between snowfields and glaciers is like the difference between a rain-drenched field and a river. A flooded field gets muddy, but its water is not going anywhere or doing anything much. A glacier is not inert. It is actively working the land around it, stacking up moraines and scratching the rock it crosses. So Agassiz could see, between the glacier he was viewing and an old moraine, patches where the glacier had sliced away the earth and left a calling card of parallel grooves, all pointing in the direction of the glacier's former flow.

Another fact evident at Chamonix, at Zermatt, and up on Les Diablerets was that glaciers are not just ice rivers topped with névé. Besides pushing moraines in front of them, they carry rocks, stones, and boulders on top and even more debris mixed in their bodies. Glaciers are too big and unstable to be just ice cubes writ large. They have crevasses that trap tumbling objects, unlucky animals, and windblown dust. During the course of a summer a boulder can travel very far on top of a glacier, until it is finally dumped onto the moraine. The boulder might travel a few miles while the face of the glacier hardly seems to budge, suggesting there is nothing trivial about that water runoff. If the glaciers quit melting, the speed of their growth might astonish the world.

Charpentier and Venetz showed all this to Agassiz and quoted the relevant passages from Saussure, then they showed Agassiz things very far from these glaciers. They took him to the unstratified mounds of the floor of the Rhône valley. Years before, Venetz had noticed these hills and asked himself what they were all about. Agassiz looked and saw what Venetz and then Charpentier had already seen. These mounds were the lateral moraines of some bygone glacier, one large enough to have filled the Valais.

At Martigny, the Rhône makes its pivot to the north, and the space between the Alpine walls becomes much narrower. Whatever catastrophe passed that way had had to squeeze through a tight space here. Rushing water would have eroded the bottom of the valley, where pressure and current are strongest, but the upper parts of the wall beyond Martigny are worn and polished—the same way glaciers scar the land up at Chamonix. Standing at Bex, Agassiz could look across

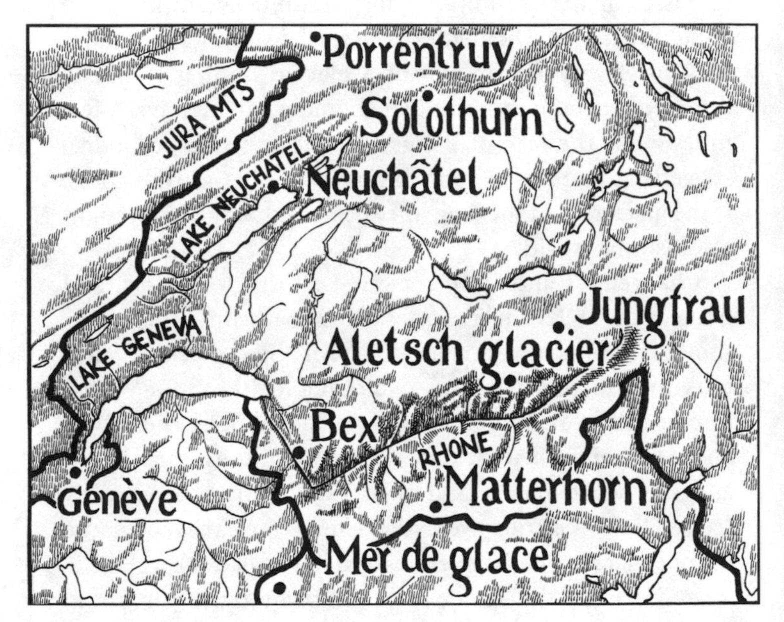

Switzerland

the valley where Charpentier pointed and see, very high up, areas along the wall that were smoothed and scraped. He had never seen a bulldozer, but he could see what had happened there. A glacier tall enough to reach high, high above the valley had sliced away protruding bits of cliff side with no more difficulty than a lion has swatting down a gazelle. In those days glaciers had outpowered mountains, and in his imagination Agassiz now understood what the valleys leading toward a bay such as Melville would be like.

Charpentier and Venetz were naturally delighted to have brought Europe's most promising young naturalist on board, but perhaps they were a bit overrun, as Agassiz wanted to learn more and more. Did they want to take a break? Agassiz wanted to go look at another mountain. Did they want a moment's quiet? Agassiz talked with the speed and enthusiasm of a crazy man, only with this exception: His sentences flowed into paragraphs, and his paragraphs made beautiful sense.

Agassiz sent off to Munich for a fellow steam engine, an old schoolmate named Karl Schimper. Like Agassiz, Schimper was enthusiastic about all of nature. By training he was a botanist, but by inclination he was curious about everything. Agassiz showed Schimper what he had seen. Bex had its mysterious potholes, the remains of waterfalls pouring off a mountain-sized glacier as it melted under a finally triumphant sun. The blocks of Monthey? Carried there by a glacier and then left behind when the glacier retreated to its little white specks high in the mountains. Schimper saw and recognized what Agassiz showed him. And Schimper matched Agassiz in enthusiasm. In November, when Agassiz at last put an end to his "summer vacation" and went home to Neuchâtel, Schimper stayed behind to continue his investigation of the Valais and the mysterious past era when the whole of the Alps had rested under a glacier.

OFF CAPE ALEXANDER,
GREENLAND, AUGUST 6, 1853...

Smith Sound spread open before the *Advance.* The waters around the travelers had been clear for days, ever since leaving Melville Bay, and now the brig was crossing the frontier of the mapped world. The atlases and globes of that era showed lines at this point vanishing into uncertainty. A light blinked on the horizon, as though some great mirror were reflecting the sun. By then everybody aboard knew that such blinking came from ice. Apparently they were not sailing into an Open Polar Sea.

They reached the ice pack the next day, and now at last they saw something more impenetrable than Melville Bay, a vast tangle of floe and icebergs, though no source for these bergs could be seen. Kane did not give up right away. His plan was still the one he had set forth before a New York audience: force his ship as deeply into the ice as he could; survive the polar winter; press on to the open sea when the sun returned. The explorers spent days searching for leads in the ice and moving forward whenever they found one.

As if the ice were not enough of a problem, the dogs drove them mad with their constant hunger. Kane complained that it was like sailing with fifty "ravenous wolves." The last of the polar bear meat shot in Melville Bay had been eaten soon after the ship entered Smith Sound. Kane balked at feeding dogs the pemmican that he had brought to keep the men alive, so besides hunting for routes deeper into the ice, the crew had to hunt food for the dogs, mountains of food. One seal did not go far among fifty and more dogs.

Kane named the anchorage point where the *Advance* came to rest Rensselaer Harbor, after his grandmother's famous New York family. It became the expedition's operational base camp for the whole of its stay. The shore was low and flat, confronting the men with none of the horror of Melville Bay's glacial wall, although well off to the southeast stood some kind of a ridge. The land, in fact, was open tundra. Grasses, mosses, and birds gave the place a feel that, if not quite Edenic, was at least not infernal. They even found a river of running freshwater.

The party began its occupation by exploring the immediate area, a chore made more difficult by the unstoppable hunger of the dogs. After one of his scouting expeditions, Kane returned to the brig with two bird's nests he hoped to carry back home as one part of a collection of natural specimens. One of the dogs immediately ate both nests. "Worse than a street of Constantinople emptied upon our decks," Kane complained to his journal.

But for all their problems, the dogs were essential. The party could expect to sail no farther. Whatever the men would find next year had to be discovered overland. If there was a polar sea open for sailing, they had to reach it by land, with the dogs pulling them and their loads. They had brought rubber boats for launching into open water.

Meanwhile, there was scouting to be done. Kane ordered his ship's surgeon and the Eskimo guide to make a long search to the south to see what lay beyond the tundra meadows. Other parties went out hunting food for the dogs and for themselves. Kane himself spent many of his days learning how to drive a dogsled. The animals were not wild, but they knew their own wills and acted as they wished unless they were under extremely firm guidance. Even tied together into a team, looking for all the world like a set of tamed, trained work animals, they would run away with the sled in any perchance direction unless the man behind them knew how to steer and how to use his whip.

The dogs were not strong enough to carry all the weight they would need to haul come spring, so it was essential to cache as much food and material as possible along the route east. That way, the dogs could carry only light loads, which would be refreshed as the party moved across country. Some of the crew would therefore have to travel and lay food depots before the long winter set in. Kane kept busy training his dogs. He concentrated especially on the ten from Newfoundland, as they were the animals closest to being tame.

Ninety miles south of the brig, the scouts of the party suddenly saw their limit. They had found the Greenland ice sheet. This was no river of ice, no glacial tongue creeping forward. It was an ice ocean, a mass 400 feet high at its border that covered Greenland from end to end and from east to west. Here was the source of Melville Bay's glaciers, as well as that of the glaciers they had seen below Upernavik and the glaciers that others had reported while exploring Greenland's eastern coast. From a rise they ascended as they approached the sheet, the scouts could see the ice surface spread to both the southern and eastern horizons. The scene was as barren as the Atlantic Ocean, and, like the Atlantic, there was no telling what mountains and valleys lay beneath its frozen waves. Before the scouts lay more ice than they had seen in all of Melville Bay. Indeed, except for perhaps one or two sailors in the Antarctic and a few Eskimos, before them lay more ice than anybody on earth had ever seen.

Back at the brig Kane was becoming more skilled at running the dogs. He could direct them easily now on a run around the ship and had passable success at greater excursions. He began taking longer runs. The animals made a strange combination of the wild and the domestic. They were still impossible to control in the presence of anything of animal or vegetable origin. They stole a cheese from the ship's store and ate some mosses that Kane had collected. Nor did they show any of a dog's familiar inclination to snuggle up against a human or do anything so affectionate as lick a hand. Yet they were not so wild that they sought out their own territory. They preferred to sleep in the brig's immediate vicinity and would have nothing to do with the doghouses the crew had made for them some little distance away.

Signs of the approaching winter were already visible. In late August the sun began to mark the days by rising and setting. By September 10, Kane noted that, except for the snowbirds, the bird life that had been so plentiful when they arrived had flown south. Even with the sun in the sky, the temperature had dropped to 14°F.

The scouts returned with the news of the ice sheet, and in the report that Hayes, the surgeon, wrote for Kane he mentioned seeing a meteor land somewhere in the ice. They could hear the collision in the distance. It had, of course, been impossible to go looking for the meteor, but it was out there, being ground up and moved around by the ice. And what would geologists say if one day, many years afterward, they found this stone and recognized its oddity? Where had it come from? How had it gotten wherever they found it?

Winter was beating down on the party, and on September 25, Kane finally sent a group to follow the coast as far as it could and establish caches for the next spring's exploration. By the end of September the sea ice around the brig had reached 14 inches.

Kane kept practicing with his dogs. He had learned the secret of the whip and could hit whichever dog he aimed at while making the whip crack loudly enough to hold the other dogs' attention. As the sled tours grew longer, flying seemed to be as much a part of them as sledding. The ice was not flat or continuous, but filled with bumps and cracks. Once, the dogs tried to jump over an especially wide chasm, missed, and fell into the water, pulling the sled in with them. Kane jumped quickly enough to save himself and then, with another crew member, rescue the dogs. Another time, the dogs leapt successfully over an abyss and brought the sled with them, but a crewman with Kane fell off and had to be rescued from the waters.

Kane was marking time now, waiting for the appearance of his eastern scouts, back from preparing the caches. There was nothing more to do except hope they had prepared well enough for everyone to survive the midday dark.

NEUCHÂTEL, FEBRUARY 15, 1837...

Agassiz's old school chum, Karl Schimper, had come over from Bex for a long visit. He read two papers on plant biology before the Neuchâtel Society of Natural Sciences and then passed out a humorous ode that he had penned and that still holds a place in scientific, if not literary, history.

The ode contained a new German word, *Eiszeit,* meaning Ice Age, and marked the first time the term had ever appeared in writing. Schimper and Agassiz had been talking, and they had concocted a new idea.

The previous November, when Agassiz returned from Bex, he had been eager to examine his own surroundings for signs of glacier actions, so he went exploring in his local hills. The Jura Mountains behind Neuchâtel are less dramatic than the Alps, being lower, less steep, and barren of glacier peaks. Their greatest claim to fame, perhaps, is that they gave their name to the Jurassic Era, when dinosaurs and primitive mammals populated the earth. The region's limestone foundation is softer than the granite Alps, but scattered along the mountainsides are erratic granite stones that range in size from that of sand grains to gravel to gigantic boulders weighing many tons.

Early in the nineteenth century Leopold von Buch examined these Jura strays and established that they came from Mont Blanc, 90 miles (90 very rugged miles) away. Von Buch made himself the foremost authority on European erratics. He showed that a series of boulders formed a ring about 200 miles in diameter, with Mont Blanc at its center. Later on von Buch established that erratic boulders in northern Europe had come from Scandinavia. How had the boulders traveled so far? Von Buch dismissed glaciers as a possibility, preferring the explanation that Mont Blanc had popped up from the earth with the speed of thunder, and, as the giant catapulted skyward, it sent its loose rock shooting farther. *Look out below, here comes 50 tons of granite.* Surprisingly, the flying stones did not make craters where they landed, nor did they shatter into a trillion splinters.

Agassiz had examined the Jura's erratics before and had even written about them, but he went for another look as soon as he got back from visiting Charpentier. Previously Agassiz had believed that these erratics had floated on icebergs to their positions, but now he saw that action was impossible. The erratics are in distinct zones, not scattered randomly as one would expect if water had covered an area. The erratic zones faced the openings of the great Alpine valleys, just where they would be if glaciers had emerged from them. A flood coming into the valley above Neuchâtel would have spread more widely as it emerged from its Alpine sluice, not run straight on to the Jura.

The Jura erratics are salted at many levels up and down the slopes. A glacier can account for this fact because it would have carried boulders on its surface and then, after melting, left the boulders to lie at whatever altitude happened to be below their resting point. Water transport, however, would favor a single level. Even an armada of icebergs, loaded with rocks, would float at the same level. If they hit the side of the Jura range, got stuck, melted, and deposited their boulder passengers where they rested, all the erratics would be lined up in a row like sand on an ancient beach.

Of course, icebergs could have melted without getting stuck and then dropped their boulders through the floods, letting them fall randomly below. But Agassiz could see that the Jura erratics have never been underwater. Their corners are too sharp. They are not rounded the way rocks, even granite ones, become eroded when they sit underwater for a time.

As he hiked over the Jura's face, Agassiz was struck by another feature of the range that geologists had previously ignored. Especially striking—now—were polished rocks on the Jura's southern slope, facing the Alps. For a stretch of over 50 miles, the Jura behind Neuchâtel show shiny surfaces with oblique furrows—that is, the grooves are perpendicular to the mountain slope. The Swiss villagers called these polished mountainsides *laves,* implying that water had washed these shiny stones. The pre-Bex Agassiz thought that the traditional explanation was generally correct. The post-Bex Agassiz, however, saw something else. Running water does not carve oblique scratches for the obvious reason that water flows down a mountain slope. A force that makes oblique scratches is not paying attention to gravity.

North of Neuchâtel is a mountain hollow called the Creux de Van. Today it is a small nature reserve that makes for a charming picnic area. It is a miniature Yosemite—a place of steep walls alongside a box-canyon valley. The floor is peppered with stray granite, and the walls include patches of *laves.* When Agassiz visited the Creux de Van, shortly after his return from Bex, he became the first naturalist, probably the first person, who could see that the hollow had been marked by a glacier.

In the Jura's interior is an area known as the Val de Ruz. Near there, at Pertuis, Agassiz saw another zone of erratics. As expected, this area faces an Alpine valley. The Aare River points at the place, but the erratics are inside the Jura. Any glacier that carried them had to move beyond the first slope of the Jura.

Agassiz saw all this with newborn eyes. The glacier markings evoked images of things unknown, of ice crawling and scratching its way overland, dragging fantastic chunks of granite with it, rubbing the land raw as it passed. Really, Agassiz needed three eyes to read glacier actions: two for close observation, and one more—a 20/20 mind's eye—to perceive the meaning of what he saw. Using all three eyes, he could look at the furrows scratched into rock polishings and see which way the lost glacier had crept. With those three eyes, he could look at a stratified mound beside a fork in a mountain road and see ancient scenery: a pool of water sitting beside the edge of a glacier. Both water and ice were now gone, but the layers of silt that had settled under the pool endured.

Within a month of his return Agassiz had seen and understood enough to believe that the Jura Mountains had once been covered by a glacier. He had already gone further than Charpentier or Venetz. Those men had conceived of a great Alpine glacier that spread from Mont Blanc and the other peaks, through the valleys, and pressed against the slopes of the Jura. Agassiz's exploration told him that the glacier had reached the interior valleys as well.

When Schimper joined Agassiz in Neuchâtel, he found that his old roommate now envisioned a superglacier covering the Alps and the Jura region. Neither of them knew anything of a Greenland ice sheet, so it was difficult to picture Alpine valley glaciers that had grown large enough to scale the Jura slopes and still press on like an army of unstoppable ants. But Agassiz believed that the glacier was greater than Charpentier or Venetz supposed.

Almost any other scientist in the world would have discouraged Agassiz. It was, they would have said, physically impossible for a glacier to go uphill. Water cannot climb a mountain, and glaciers do not flow up slopes. But Schimper encouraged Agassiz. In 1835, the summer before his visit to Bex, Schimper had examined erratics in Bavaria and concluded that they had been carried by a glacier that had come down from the north. When Agassiz told him of a Jura glacier, Schimper wrote home to Munich for a notebook in which he had explored ideas on a Bavarian glacier.

Lightning seems to have jumped between the two old friends, and suddenly they produced a new idea: The Bavarian glacier, the Jura glacier, the Alpine glacier, and the north European glacier that had carted von Buch's erratics from Scandinavia were all the same glacier. There had been a time in the geologically recent past when the northern hemisphere from the Arctic Ocean to the Mediterranean Sea had been covered by one enormous sheet of ice. The mystery of human imagination showed itself in that lightning bolt—ignorance went in, a new idea came out.

By 1840, Agassiz and Schimper would become jealous rivals, each accusing the other of stealing the Ice Age idea. Historians have felt it their duty to try to sort out who should receive the credit, although they usually agree with the geologist-historian Albert Carozzi that it is "impossible at present to outline their respective parts in the fundamental concept of an Ice Age." But surely the truth is that neither could have done it alone. The idea is the joint fruit of two minds, not one.

Agassiz was a fact man who rarely connected the dots. His theories tended to be constricted by an unwillingness to push to a conclusion that was even one leap ahead of the facts. Years later William James recalled a trip he took as a young man with Louis Agassiz, by then an old man. James said that Agassiz insisted on amassing facts and voiced a frequent contempt for James's fondness for jumping to conclusions. It is difficult to conceive of such a fact-bound soul leaping to the century's most amazing geological conclusion without a Schimper to urge him on.

Schimper, for his part, was a theorist, a generalizer who hoped to create a geometric model of plant structure. He was more interested in the big picture than in the details; and, on his own, he was unlikely to consider the many fine points of glacial motion and melting, of climate and terrain, that could translate his old ice from a white, geometric shape into a living glacier. Together, however, the collaborators enjoyed a respect for facts and an ability to generalize that could produce a real theory of an Ice Age.

Behind the teamwork lay a common enthusiasm for the meaning of their work and their schoolboy habit of talking about what they were learning. Agassiz and Schimper had schooled together in Heidelberg and then transferred together to Munich. They had roomed together and given each other nicknames. Schimper the botanist, was called "Rhubarb," while Agassiz, already a fish enthusiast, was "the Carp." At Neuchâtel, Rhubarb and the Carp were again together, again with new knowledge, and again able to say to themselves what they would not say to others. The baby born of these sessions was the original version of the Ice Age theory.

Today's standard theory of the last Ice Age only loosely resembles the one worked out in the early winter of 1837. Most notably, current theories do not link the Swiss ice sheet to the northern European glacier. Those bodies are now said to have been two separate ice sheets with a few hundred miles between them. But the Agassiz-Schimper theory had a meaning that satisfied its creators more than the modern theory could.

Schimper was an heir to Germany's romantic science, the *Naturphilosophie* that Cuvier had scorned in his last lecture. He viewed biological change as progress toward a single ideal type—man. One of the leaders in the movement, Lorenz Oken, and a teacher of both Agassiz and Schimper, had written that man was the goal of nature. "Animals are only the persistent fetal stages of man," he said. Schimper's search for the ideal plant form was a botanical relative of Oken's interest in the ideal animal form.

Naturphilosophie was too focused on general forms to be deeply concerned with the objective machinery that progresses toward the ideal, but it welcomed any proof of the reality of the process. Fossil evidence, for example, plainly showed that animals' forms had become increasingly "modern" as they came closer to our own time. Now the Ice Age theory provided a revolutionary engine for wiping out the old life-forms. When the ice retreated, newer, more ideal groups of life appeared.

Agassiz knew the *Naturphilosophie* theories, but they were too abstract and mystical for him. They took "the forces of history" literally and required some kind of active historical spirit. For Agassiz, Ice Ages were evidence of a providential God working out a plan to create Man in His image. Agassiz also saw Ice Ages as a way of proving conclusively that Cuvier had been right and Lamarck wrong. An Ice Age that wiped out all life on earth would make it impossible for today's creatures to be the blood descendants of ancient species.

These philosophical justifications of an Ice Age strike contemporary readers as bizarre and archaic, and it may seem paradoxical that an idea as good as that of the Ice Age would rest on such unscientific and personal motives; but reason cannot get beyond its own ignorance. New ideas come from acts of imagination, from guesses that inspire further inquiry because they address private anxieties.

All those personal, irrational interests kept Rhubarb and the Carp up late at night talking about a great glacier that swept down from the Arctic to erase and replace the world's fauna and flora. In their bull sessions they worked out many details. If the world was once covered by a glacier, it must have been colder then, but almost every geologist at the time believed instead that the world had been hotter and was cooling down from a fiery birth. A few geologists—Lyell was the most prominent—believed the earth's overall climate had been steady, though he granted local fluctuations. None imagined that the world had been colder and was now thawing out.

Yet for Rhubarb and the Carp, cold snaps were acceptable. Agassiz remembered those frozen mammoths that Cuvier had analyzed. They could have been trapped in the great frost at the start of the Ice Age. Schimper knew that Germany's greatest poet, Johann Wolfgang von Goethe, had speculated in 1829 on a cold period of ice and water. "We say there was an epoch of great cold," Goethe wrote, "at a time when waters covered the continent up to 1,000 feet in height [and]...the glaciers of the Savoy mountains went farther down, all the way to the sea."

So the two friends did not have to be completely idiosyncratic to believe in a past cold. They worked out a theory that said the world had been hotter (as most agreed), but as it cooled, it sometimes fell into deep chills and then warmed up again, although not reaching the pre–Ice Age warmths. Schimper, the generalizer, drew a simple little graph to illustrate the temperature curve over geologic time:

Another problem was that the great glacier had come down from the Arctic, but the Jura erratics had come from Mont Blanc and could not have simply been picked up during the glacier's travels from Norway to Neuchâtel. Agassiz found a solution in theories of mountain building. Although Lyell believed that mountains like the Alps rose very slowly, the best continental geologists still favored catastrophic origins. Von Buch, in Germany, and a geologist named Élie de Beaumont, in France, argued for the sudden rise of mountains from the earth. Agassiz and Schimper took their side, proposing that during the Ice Age, when Europe sat under a glacier, the Alps had suddenly been born, climbing to their heights like dandelions that grow in a day. As Mont Blanc stabbed its summit above the great glacier, huge chunks of it were ripped off and came skidding down the ice like skiers tumbling down slopes. Some slipped as far as the Jura, where they remain today.

It takes an especially nonjudgmental reader to think that this "explanation" was anything but a laughably thin straw. Yet by hanging on to that will-o'-the-wisp, the friends kept the one-great-glacier idea afloat. Otherwise, they would have had to admit that the facts of the case would never allow for a time when a single sheet of ice had wiped out the creatures of the northern world and carried the erratics from Mont Blanc. Error was replacing ignorance.

By February 15, 1837, the two had worked out the principles and meanings of the one-great-glacier theory. Schimper celebrated by writing an ode, titled "Die Eiszeit," which he passed out to Agassiz's most devoted admirers in Neuchâtel. It was the pair's first effort to spread the Ice Age idea beyond the view of its two composers. Years later, when Agassiz wrote with only bitterness about his old friend, he dismissed Schimper as "a man who wants to present new laws of experience through an ode and other fantasies of the mind." Agassiz—ignoring the secret that he had just gone through the most irrational and creative enthusiasm of his life—was determined to make a rational case for the Ice Age.

LONDON, MARCH 19, 1837...

Lyell's conversations were not as intense that winter as the talk at the Agassiz-Schimper thinkathon, but neither were his remarks limited to *Please, pass the salt.* In a letter to his sister Sophy, Lyell reported a conversation that exposed the state of mainstream geology as the two young men in Neuchâtel prepared to startle the world. The geology that Lyell knew had rattled tradition, but it had not yet overturned it.

Lyell told his sister that he had met with Henry Fox, Lord Holland, an old Whig with two titles. Absurdly, he was the Baron Holland of Foxley and the Baron Holland of Holland, giving him a name ripe for satire, yet he was a serious man and one more radical than Lyell could ever be. Thirty years earlier Lord Holland had been "Lord Privy Seal" in the government that abolished the slave trade in the British colonies. He had been a pro-tolerance, anti-Tory politician his whole life, a man of wide learning and modern views who felt able to hold his own when talking to Lyell about geology.

The link that season between a specialist like Lyell and a layman of Lord Holland's stripe came from William Buckland's new book, *Geology and Mineralogy, Considered with Reference to Natural Theology.* The book was a compendium of the latest geological knowledge, organized to show how geologists could see God's plan at work over the eons. It was one in a series of treatises financed by the Earl of Bridgewater to show "the power and wisdom of God as manifested in the creation." The book was a great success, and naturally people wondered what Lyell thought of Buckland's Bridgewater.

Lyell did not like "natural theology," but he launched into a solid defense of Buckland's discussion of fossils. Fossils denied the tradition that the earth was only a few thousand years old. Buckland reported, and Lyell agreed, that the "terraqueous globe" had seen many revolutions in life-forms.

It may seem a little surprising that Lyell went along with Buckland's view. Lyell was anticatastrophe, yet he was praising a book that referred to "the sudden disappearance of a large number of the species of terrestrial quadrupeds" and "the catastrophe by which they were extirpated," sounding exactly like Cuvier. But catastrophes were a detail in the larger vision that Lyell, Buckland, Agassiz, and Schimper all shared: The study of fossils provided decisive evidence that an enormous amount of time and biological activity had preceded the appearance of any humans.

The conversation with Lord Holland then led, Lyell told his sister, "to a talk on new species, and that mystery of mysteries, the creation of man." Science historians pop to attention when they read that passage because two years later Charles Darwin published a book about his around-the-world voyage aboard the *Beagle* that mentioned "that mystery of mysteries—the first appearance of new beings on this earth." Lyell's use of the phrase two years before Darwin waves at us like a red cape at a bull, and just like that red cape, it diverts us from what really matters. For if we look above the fluttering distraction, we can see what was radically threatening about the Ice Age theory. Naturalists and their readers alike had their eyes focused on the fossil coin's head side: the appearance of new species on this earth. Agassiz and Schimper had flipped the coin over to examine its tail: the disappearance of old species.

This is just one more difference between that time and today. For us, the mystery of mysteries about, say, the dinosaurs is what became of them. Was it a great meteor or something else that wiped them out? In Lyell's day the mystery focused on the other end of the story. Where did the dinosaurs come from? Agassiz, probably without noticing and surely without intending, was about to start shifting attention to the destructive side of the coin.

Liberal Christians who accepted geology and still insisted that the Bible had something to tell us about creation found that geology supported this basic theme of "the Inspired Narrative": The one living God had made the world according to a single plan, had made it out of love, and had interfered providentially in its history. The geological evidence for all this doctrine rested heavily on the constant appearance of new species.

New species showed that the Creator was still interested in what happened on earth. The splendid adaptation of each new species to its circumstances proved God's loving attention to His creatures' needs, as did the overall globe's general adaptation to serve life. The recurring design elements in species after species proved a single creative power had created these animals and plans according to a consistent purpose. In short, the whole justification rested on creation and presented God as the world's "common Author and Preserver." God the Destroyer has no place in this theology, but an Ice Age theory brings destruction front and center.

As Lyell and Lord Holland exchanged words, they remained focused on the creation side of their coin, not even realizing they were ignoring half the story. Holland responded to Lyell's "mystery of mysteries" by snorting that the present generation of scientists had come no further toward explaining new species than Lucretius, the Roman poet and Caesar's contemporary. Lord Holland added that he could have taken passages from Lucretius and turned them into "mottoes" for each of Lyell's three volumes. By citing Lucretius, Rome's most notorious atheist and mocker of the gods, Holland surely felt like a scandalous old Whig, one faithful to the religious radicalism and toleration that had marked his whole career. But today's reader wants to reach back through the calendar, grab these people by the shoulders, and demand, *But what about the destruction? Was the same designer also doing all the rubbing out?*

The Ice Age theory raised that question directly. Catastrophic floods, even if they were not Noah's flood, had a certain biblical ring to them, but destruction by ice had no such tradition behind it. By focusing on the machinery of destruction, Agassiz would be trying to change the discussion from birth to that other mystery of mysteries, death.

RENSSELAER HARBOR,
GREENLAND, OCTOBER 3, 1853...
Probably only one more week of sunrises remained before the midday dark would begin. Astronomically, sunrises should have persisted until October 24, but a mountain ridge blocked the narrow section of sky where the sun should appear above the party's base camp, and Kane calculated that their long test would begin on October 10.

The landscape was already changing. An "ice foot" had appeared. This curiosity is an ice fretwork that forms where the frozen sea meets the land. The sea, even when frozen, still moves with the tide, creating a region that is sometimes frozen sea and sometimes frozen land. As the tide comes in, it forms a kind of "foot" that builds up, twists, melts, and refreezes, giving the harbor an arabesque border between sea and land. Already the foot was many inches thick and over 30 feet wide.

The group that Kane had sent out to cache supplies for the spring returned with news that its way had been blocked by an enormous glacier. The scouts did not fully understand the magnitude of what they had found, but they grasped enough to see that this was no route to the north. They had tried to go around the glacier, but the sea ice in that region was unstable. The milky way of icebergs that reached out from the glacier's face formed an ice maze that looked as impenetrable as an equatorial jungle. They had discovered what Kane would name the Humboldt glacier, the largest glacier any of them had ever seen. At the close of the twentieth century, the Humboldt is still the largest known glacier on earth, spanning 60 miles at its snout. The route north was blocked on Kane's side of Smith Sound. In the spring they would first have to cross the water and then go north, hoping that nothing so dramatic blocked that route.

Passable or not, they were up there for the winter. The sun set and did not rise. At first a predawn light could be seen as the sun came close to the horizon without soaring above it, but then that reminder of more southerly latitudes also disappeared. The party got a reprieve of sorts at the end of October when the moon came up and stayed there. Like the sun, the moon does not hover along the equator but travels up and down across the tropical latitudes. The moon even reaches a bit above the sun's highest point, climbing to 25° N, which was where it stood in October 1853. Kane's group had eight days of constant moonlight, and then it too went away, setting and slipping farther south.

There was nothing to do now but read, work on the lamps and fires, study the stars, and try not to go mad. On December 12, the men turned out to watch the planet Saturn disappear behind the moon. An event of that sort was fully predictable in astronomical data and gave the group confidence in its location, 78°37' N and 70°46' W, 780 miles from the North Pole. Then Saturn reappeared, and the party returned to the monotony of freezing in darkness.

One detail that Kane had not foreseen was the effect of the long night on the dogs. They could not stand it and

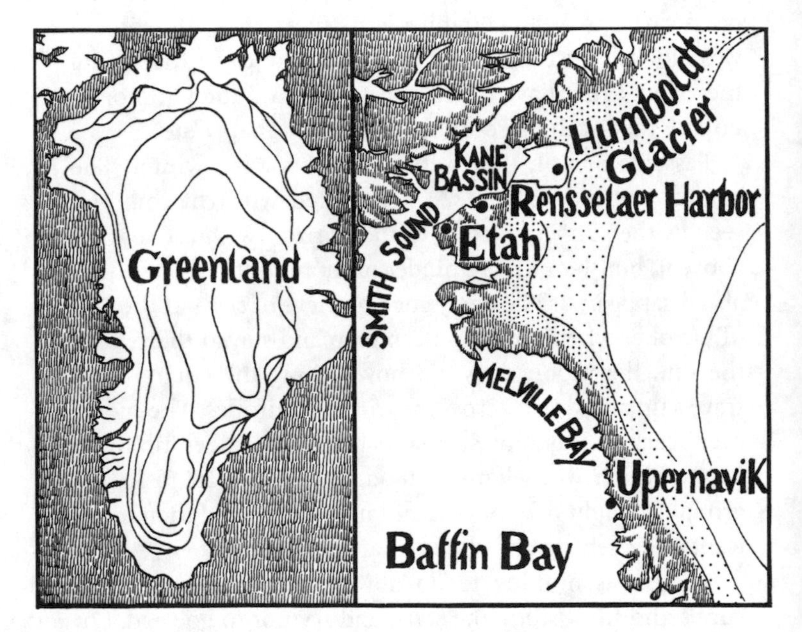
Greenland
KANE
BASSIN
Humboldt
Glacier
Rensselaer Harbor
SMITH SOUND
Etah
MELVILLE BAY
Upernavik
Baffin Bay

Greenland

became as depressed as the most suicidal human. Some like Old Grim, the expedition's lead dog, simply wandered off into the night. Others died from what Kane, with his medical training, described as "an anomalous form of disease, to which, I am satisfied, the absence of light contributed as much as the cold." He added:

> *Their disease is as clearly mental as in the case of any human being. The more material functions of the poor brutes go on without interruption: they eat voraciously, retain the strength and sleep well. But all the indications beyond this go to prove...true lunacy. They bark frenzied at nothing, and walk in straight and curved lines with anxious and unwearying perseveranc....Sometimes they remain for hours in moody silence, and then start off howling, as if pursued, and run up and down for hours.*

Because of the dogs' importance to the project, this unexpected development was taken seriously. The dogs received great attention, and even though their presence made life difficult for the men, they were brought below deck. There the crew petted them, cleansed them, and gave them whatever comfort oil lamps could provide. Yet the dogs were dying. One by one, they would go into spasms and become lethargic. Typically, they died less than two days after the onset of spasms.

Almost all of these dogs had been born and raised above the Arctic Circle, and at first the party could not believe that the night would drive them to their deaths. But by January 25, with still another month before the sun's return, Kane had given up any hope of saving them. Besides being terrible to watch, the destruction of the dogs put an end to

Kane's plans for the coming spring and summer. Even if a few dogs did survive, there would no longer be enough of them to carry everybody overland in their quest for a polar sea. At best, they could send out a few more scouting parties and fill up a few more spots on the map, but any hope of actually using their harbor as a launching base for a trek to find the Franklin party was gone. (Although, of course, it was now obvious to everyone that none in the Franklin group could possibly have survived the eight Arctic winters that had passed since they were last seen.)

In early February the weather turned steadily colder. The party's thermometers showed, as a high, –60°F and a low of –75°F. And the days stayed that way. When, at the very end of February, the sun again appeared above the horizon, they had endured 140 days between sunset and sunrise.

They could at last see Rensselaer Harbor again. It had become an ice wilderness. Especially dramatic was the ice foot that, in the darkness, had grown to a wall 30 feet high and 120 feet wide. Plainly, it would be well into summer before all that ice went away, if it ever did.

The persistence of terrible cold into March took Kane by surprise. Even with sunlight, the days were –46°F. Kane ruefully understood what his thermometers showed. There was no sign of temperatures letting up at these extreme latitudes. The theory of an Open Polar Sea had required that, by March 21, when the sun crossed the equator, the latitudes at 80° N and above should be warmer than temperatures farther down the Greenland coast, in, say, Upernavik. But the cold was not diminishing as they approached the Pole. None of the plans or theories they had brought with them had survived the winter.

NEUCHÂTEL, SWITZERLAND, JULY 24, 1837... Agassiz approached the annual meeting of the Swiss Society of Natural Sciences with great plans. As that year's president, he called to order a room full of distinguished scholars. Even the great Leopold von Buch had come from Berlin to hear Agassiz deliver his latest paper on fossil fish. France's leading geologist, Élie de Beaumont, was delayed, but he too was expected to grace the meeting. Young naturalists of strong promise had come as well. Amanz Gressly, for example, only twenty-three years old, had a paper with him that should complement Agassiz's fossil study quite nicely. He planned to introduce an idea about changes in ancient environments that explained why fossils were distributed unevenly through a single geological stratum.

The audience also looked forward to refreshing old acquaintances and starting new ones. Jean de Charpentier was there, of course, and besides seeing his friend and former student presiding over the assembly, he could see his old schoolmate, von Buch. Meanwhile, young Gressly could hope to impress the greats around him. The meeting's agenda scheduled two days of papers and discussions to be followed by walking in the Jura Mountains, where everyone could chat about the work of the meeting.

From the front of the room Agassiz welcomed the assembled savants and began a presentation remembered as the Neuchâtel Discourse. At its opening Agassiz boasted that this Swiss group, though much smaller than similar British, German, and French scientific societies, was older than any of them and had been their model. Then suddenly Agassiz's chairmanly boasting took an obscure turn as he said, "Just recently, two of our colleagues have generated through their investigations a controversy of far-reaching consequences for the present and the future."

Who were these two colleagues? Agassiz did not immediately clarify whom he meant, and the audience—anticipating a lecture on fossil fish—could not fairly be expected to recognize so vague a reference to the glacial research of Ignace Venetz and Jean de Charpentier. It took another few words before the audience began to realize that Agassiz was not speaking about fish at all. In fact, Agassiz had been up all night writing a summary of his glacial work. Most people in the room had had no notion of Agassiz's latest interest and could not have been more surprised if he had burst into a Rossini bel canto. Especially alarmed was von Buch, who had journeyed to the provinces only to find himself ambushed over his own expertise. He had been studying the Alps since before Agassiz was born; and twenty-six years previously, in the classic report that showed that the Jura's granite blocks came from Mont Blanc, von Buch had specifically rejected the notion that glaciers might have moved the boulders. Now Agassiz was trying to call that old issue back from its tomb, and von Buch took this return of the glacial theory as a personal insult. Even months later he would still be in a fury about it. Five months after this meeting Humboldt wrote Agassiz that von Buch was still raging, considering, as he did, the subject of erratic boulders "his exclusive property."

Meanwhile, Charpentier could only be pleased by this opening, which showed how deep Agassiz's conversion to the glacial theory had been. And Agassiz admitted that this first part of his discourse was simply a restatement of the ideas of Venetz and Charpentier.

Most in the audience, of course, were not so personally committed to one side or the other of the science. Yet they too were surprised by Agassiz's declaration that he had joined the glacialist camp, and they reacted the way an audience of eminent Catholic priests might act if their presiding bishop began his welcoming address by presenting the case for some Protestant doctrine. Agassiz's audience became noisily uncomfortable, stirring restlessly and breathing hard. Charpentier was distressed by the depth of ignorance about glaciers that the audience members continued to show. They had learned nothing in the three years since he had given his own glacier paper before this same society.

The hostility to Agassiz's discourse is not easy for modern readers to understand. He assembled the facts logically and based every assertion on visible evidence. The Neuchâtel naturalists had a difficulty we no longer experience in imagining great ice. Why, we wonder, should anybody balk at the idea that the ancient moraines of the Rhône valley were built when the Swiss glaciers were much larger than today? Who could pooh-pooh Agassiz's comment that far from the present glaciers they could find "moraines perfectly similar to those which still surround glaciers"? Yet balk and scoff the Neuchâtel savants did.

Agassiz found the nerve to continue. He began discussing the evidence he had noticed in the Jura. Again, his argument was based on observation. He contrasted the great boulders with the smaller debris beneath them. The erratic boulders showed almost no signs of erosion and still had sharp edges, but they rested on top of small stones and grit that had been worn into round balls. How to explain this sight? He said that if water had carried all those stones, the smaller rocks would have settled on top of the erratics, not on the bottom, and he then offered a simple—to today's ears—explanation. The small stones and grit had been dragged beneath the glacier and sandpapered into balls. The sharp-edged boulders had ridden on top of the glaciers, where they escaped erosion. As the ice melted away, the uneroded boulders descended, finally settling on the sharply eroded debris at the bottom of the glacier.

With this argument Agassiz had moved to what we might call "stage two" of the glacial theory. Stage one had been endorsed by Charpentier and claimed that the glaciers now visible in Switzerland had once been much larger. A stage-two theory—what Agassiz said about the Jura—argued that some places had once been covered by ice, which had since completely melted away. If the audience refused to accept the stage-one idea—which to us sounds as foolish as rejecting the idea that a lake might have once been larger—it could not have been expected to accept stage two, the equivalent of believing that a dry valley had once held a mighty lake.

Having failed to persuade his listeners of anything, Agassiz then pushed them beyond endurance by offering still another wrinkle to the glacial theory—call it the "stage-three theory": Ice once stretched from the North Pole to the Mediterranean, burying the whole northern world beneath one gigantic glacier. This idea shocked everybody in the room, including Charpentier. Von Buch called out sarcastically, in Latin, "Saint Saussure, pray for us."

Catcalls like this were almost unheard of at such meetings, but all pretense of scholarly decorum had disappeared. The gathering was taking on the feel of a political rally where a poorly received candidate refuses to sit down.

People shouted angry questions from the floor, accompanied by groans and ruder noises.

Agassiz pressed on. His arguments in support of an Ice Age depended less on observation and more on imagery as he asserted that, "A Siberian winter established itself for a while on ground previously covered by a rich vegetation and inhabited by great mammals, similar to those now living in the warm regions of India and Africa. Death enveloped all nature in a shroud." Several times he referred to Cuvier's old example of the frozen mammoths, assuming they had been warm-weather creatures caught in a sudden refrigeration.

Agassiz did have one argument that, in calmer circumstances, might have intrigued at least some of the geologists in the room. "I have no doubts," Agassiz said, "that most of the phenomena attributed to great diluvial currents . . . have been produced by ice." But by then his audience was in no shape for intriguing speculation. It was loud and getting louder. Von Buch was turning into an angry Robespierre, leading shouts against a counterrevolutionary.

Von Buch's present-day defenders point out that Agassiz did have one plainly cockeyed theory in this part of his discourse, his claim that the Alps had risen only after the Ice Age had begun. But von Buch and the audience were already out of control before Agassiz presented this notion. They had become restless during the logical, observation-based part of the lecture. The speculative, metaphorical climax only made them more unruly.

And when the postlecture discussion began, the scene became riotous. It turned so vehement that some observers feared the introduction of fisticuffs. Von Buch had many hostile questions. Agassiz held his own, facing the questions and citing the facts that supported his case. He showed anger, but the emotion that lasted was dismay. He would not soon forget how his colleagues, in their eagerness to dismiss his conclusions, could be so ready to wave aside facts.

Even today the scene at Neuchâtel makes many historians of science uncomfortable. It grossly contradicts the pleasing belief that science progresses on an accumulation of facts and rational argument about what the facts mean. Apologists for that creed look for mitigation in the Neuchâtel scandal. They want Agassiz to have been a little short on facts and his audience to have been rationally skeptical. But Agassiz had ample facts while his detractors rode on an emotion with only the scantiest of reason's excuses.

The scandal of the Neuchâtel Discourse showed how unstable a foundation Agassiz had to build on. His tone and manner took the fundamentals of glaciers for granted, but his audience was profoundly ignorant of those basics. Worse, this surface ignorance rested on a deeper level of confusion. Charpentier said the dispute that day "had no single result because there was no agreement on the principles." When challenged directly and in such a complex way, each scholar fell back on his own idiosyncratic theory of the earth. Agassiz's discourse had widened the theoretical fissures among geologists. Besides dividing the glacialists like Charpentier from the antiglacialists, he had split those who thought it possible that the Alps had sprung up more or less instantaneously from those who thought mountain building was a long, slow process. He had also widened the split between those who thought the earth was still cooling from a primeval, fiery birth and those who favored steady-state or more fluctuating temperatures. Then Agassiz gave the

Lamarkians and Cuvierians in the audience something to fight about as well. He concluded his presentation by saying that, thanks to the mass extinctions brought on by the Ice Age, "there is complete separation between [the species of] the present creation and the preceding ones."

Yet even geology's immaturity cannot explain the depth and unanimity of the reaction against Agassiz. There was a third level of instability under Agassiz's feet. He had hit some unvoiced dread, against which facts and theories were no weapons at all. Ever since Newton, science had been reinterpreting the central doctrine of all those Christians, Muslims, and Jews who worship the God of Abraham: the belief that God acts in history according to a purpose. For Agassiz, the Ice Age seemed to reinforce his belief that Providence directed events. God had miraculously intervened in nature's drama by suddenly drawing down the curtain on one act, wiping out all its players and introducing a whole new set of performers (starring ourselves) in its final act. But for most of the audience, Agassiz expounded a shocking vision of nature, one in which nature itself could become the enemy of life. To many ears, that idea suggested a world more indifferent than providential, and they feared the arbitrariness of such a place.

Mired in such a terror, the disputes became so intemperate that the audience's shyer members decided to lie low. Amanz Gressly nervously kept his paper to himself. Thus, his article—destined to become one of ecology's founding documents—is dated 1838 instead of 1837.

The next day Élie de Beaumont arrived in Neuchâtel to find the meeting was still in a hubbub. He soon joined the side of the antiglacialists. Agassiz tried to rescue his position by reading a supporting message from Karl Schimper, but there was no authority in the world who could have saved Agassiz's effort that day.

Nor was the third day any better. The assembly left before dawn for its outing in the Jura. The leading scientists in the group—Agassiz, von Buch, and Élie de Beaumont—rode together in a carriage drawn by four white horses, but they were in no mood for easy conversation. One local diarist described the field trip:

> *In general, I was convinced by my short acquaintance with the leading scientists of the party that a great amount of jealousy and egoism existed between them. Élie de Beaumont was, during the entire trip, as cold as ice. Leopold von Buch was walking straight ahead, eyes on the ground, mumbling against an Englishman who was talking to Élie de Beaumont on the Pyrenees while we were in the Jura, and complaining rather offensively about the stupid remarks made by some amateurs who had joined the group. Agassiz, who was probably still bitter about the sharp criticisms made by von Buch of his glacial hypothesis, left the group immediately after departure and was walking a quarter of a league ahead all by himself. . . . Apparently everyone wanted to be left alone with his own silent thoughts.*

Only three days earlier Agassiz had seemed the most capable and accomplished naturalist of his generation. Now he had created a scandal, and his soundness could no longer be taken for granted. If today he could propose an Ice Age, what nonsense might he endorse tomorrow?

ABOARD A RHINE RIVER STEAMER,
GERMANY, AUGUST 29, 1837...

A month after the Neuchâtel scandal Lyell seems not to have heard gossip about it. Or if he had, he ignored it. He wrote a letter to Charles Darwin while his boat steamed through one of Europe's most splendid regions. The Rhine is lined with a high escarpment, a long scar that looks dramatic to even the most casual traveler. For a geologist with Lyell's eye, the valley wall provided a long cutaway view of the earth filled with eye-catching details to entertain his brain. A string of castles sits along the cliff top. Lyell, no doubt, glanced up at these sights as he worked on his letter. He was reaffirming, for Darwin's sake, his explanation for the source of erratic material, and he did not bother to reject the new glacier explanation.

While Agassiz was giving his Neuchâtel Discourse, Lyell had been up in Norway, hammering on rocks and shaking his head over Norway's political trends. He told his sister he had been saddened to encounter "a democratic system pushed so far as it is at present, so that representation is virtually in the hands of peasants or small yeomen." To Darwin he wrote more scientific observations. Stone from Norway was scattered all over northern Europe, and he described its route: "The blocks of red syenitic granite, which I hammered away at in Norway, and which I saw there *in situ* ...have been carried with small gravel of the same, by ice of course, over the south of Norway."

Darwin, upon receipt of the letter, understood that by "ice of course," Lyell meant icebergs, not glaciers. Darwin had seen just such transport during his cruise on the *Beagle.*

In Eyere's Sound, Chile, he had spotted icebergs "loaded with blocks of no inconsiderable size, of granite and other rocks, different from the clay-slate of the surrounding mountains." So there could be no doubt that icebergs do carry rocks taken from deep inland.

Lyell continued with his letter, describing the route of Norway's red granite: "down by the south-west of Sweden, and all over Jutland [in Denmark] and Holstein [further south] down to the Elbe [the German river at the base of the Danish peninsula; the shortest distance between southernmost Norway and the Elbe is 300 miles]." He then traced the granite's path down to the next important river beyond the Elbe, the Weser, and then still farther south. This area—containing the Elbe, the Weser, and other rivers—is the classic flatland of northern Germany. Military historians know this region as a highway for armies going in or coming out of Germany. Lyell, as von Buch before him, noted that the whole plain was salted with Scandinavian stone.

Von Buch believed that the stones had been thrown by a catastrophic explosion, so the point where the erratics ceased to appear implied nothing more than that the areas south of the erratics were beyond the range of the catastrophe's force. Lyell, however, did not believe in catastrophes. When he told Darwin that the stone had been carried "by ice of course," he meant it had been floated there, not blasted there by a sudden, unheard of event. For him, the disappearance of erratic material marked the end of the area that had seen floating icebergs.

Lyell traced erratics almost as far as the cities of Münster and Osnabrück, two points not far from the German-Dutch border. Just north of Osnabrück lie some low hills that mark the base of the northern plain. Lyell told Darwin, "it is curious that about Münster and Osnabrück, the low secondary mountains have stopped them [the erratics and the icebergs that carried them]; hills of chalk...&c. which rise a few hundred feet above the plain of north and north-west Germany, effectually arrest their passage. This then was already dry land when Holstein and all from the Baltic as far as Osnabrück or the Teutoburger hills was submerged." Today's geologists would say that Lyell had spotted the southern location of one of the Ice Age glaciers.

Lyell had just laid out for Darwin the countertheory to the Ice Age. It was a kind of Noah's flood in slow motion. The diluvium, as Buckland called it, or drift, as Lyell and his colleagues preferred, was the result of an inundation from the sea. Instead of coming in forty days and forty nights, the process had been much slower, but the result was the transporting of the same materials once cited as proof of the biblical deluge. The drift theory shows what a radical step Agassiz had taken in extending Charpentier's idea to a general Ice Age. It was possible to conclude that the Alps had seen a great glacier and still believe that iceberg drifts accounted for the more northerly effects. There were other explanations for the erratics' origins as well, but for years to come the chief rival to Agassiz's Ice Age theory would be Lyell's drift theory.

Part III

Changes of Heart

BEFORE THE FACE OF THE HUMBOLDT GLACIER, GREENLAND, APRIL 27, 1854...

As soon as he could, Kane journeyed to see the new glacier that his scouts had found the previous autumn. Kane wrote in his journal that he had secretly thought there must be something like an ice river at Greenland's northern end, and yet when at last he found it, he could hardly believe it.

Kane and one companion had climbed a cliff to obtain a clear look at the glacier, the face of which extended to the horizon and beyond. If such things can be dated, this was the moment when great ice at last became coherent in the human imagination. Kane saw both the ice in front of him and the forces that had put it there. He expressed himself in the style of romantic learning that would excite his contemporaries, imagining an Australia-sized continent

> *occupied through nearly its whole extent by a deep unbroken sea of ice that gathers perennial increase from the watershed of vast snow-covered mountains and all the precipitations of the atmosphere upon its own surface. Imagine this moving onwards like a great glacial river seeking outlets at every fjord and valley, rolling icy cataracts into the*

> *Atlantic and Greenland seas; and having at last reached the northern limit of the land that has borne it up, pouring out a mighty frozen torrent into unknown Arctic space.... I was looking upon the counterpart of the great river system of Arctic Asia and America. Yet here were no water feeders from the south. Every particle of moisture had its origin within the Polar circle, and had been converted into ice. ...Here was a plastic, moving, semi-solid mass, obliterating life, swallowing rocks and islands, and ploughing its way with irresistible march through the crust of an investing sea.*

Kane's romanticism was not just a pose. He was such a romantic that instead of cursing the ice for barring his way north, he almost fell down and worshipped it. Then he made his way back to the difficulties of Rensselaer Harbor.

The party's situation back at the brig was desperate. The men were broken, more aged and weakened by the winter than anyone had predicted. Earlier in April one of the crew, Jefferson Baker, a boyhood friend of Kane who had come north for auld lang syne, died suddenly, terribly, in the manner of the suffering dogs. He was seized with spasms, went into a kind of lockjaw, and was gone the next day.

The plan to press on to the north in search of passage was gone too. The men were too disheartened, only a few dogs were still alive, and, as they soon discovered, the caches set out the previous autumn had not survived the winter. Polar bears had found and eaten them all. Kane had sent out a party in mid-March, this one to head north and avoid the great glacier. The group had made a commendable try and had engaged in the kind of determined struggle that would win the scouts stirring chapters in boys' books, but they returned to the brig with nothing accomplished.

Another surprise had been the sudden presence of people. Unknown to the party, an Eskimo village lay at Etah, about 70 miles off to the southeast. The villagers appeared outside the ship one day and were invited on board. Almost immediately one of the Eskimo visitors destroyed the party's rubber boat. He sliced it up for its wooden slats and put an end to any further fantasies about launching anything into an Open Polar Sea.

Kane's revised plan was to wait for the return of summer and the reopening of the water to sail back home. In the meantime, his men would scout the area as well as possible and push back the map's frontiers a little farther. Modest though it was, the strategy looked good. Already the snowbirds that had gone south in early November returned on May Day.

Then a second crewman, Peter Schubert, fell sick and died. Kane too became miserably ill. It was something of a miracle that his heart had not killed him long before this moment. He became delirious and showed too many of the symptoms that had killed the dogs. The men expected him to die and did not much regret it. Kane had led them to disaster for nothing in return; however, he lasted beyond the normal two days, then he lasted two weeks, and finally it seemed he was not going to die after all. Equally promising, the party could see signs of progress in escaping the ice harbor. The snow had disappeared from around the brig, and the floes were beginning to break apart.

Scouting parties continued. Another group tried to explore the northern side of the channel but found it extremely difficult. The party did set the northernmost point reached by the Kane expedition or any other Greenland explorer to that date, 79°45' N and 69°12' W, but the snow up there was too deep and the hunting too scarce for the party to survive, so it returned to the brig.

Kane decided to try one more push, this time along the channel's southern coast, forcing passage beyond the great glacier. He sent two parties out. The first, led by Kane's trusted right hand, James McGary, reached the face of the Humboldt glacier on June 16. They obtained an excellent view of the glacier, advancing to within 200 yards of its front, but the sea was filled with barricades, ice mountains with peaks behind peaks behind peaks. They could see no way through and turned back.

A second group tried two days later. This team of two men, Hans Christian and William Morton, used a dog team to travel farther out into the frozen sea. The icebergs were piled so close together that sometimes they could not ride between them and had to carry the sled, twisting it through the narrow gaps. The siege of icebergs limited their view to perhaps 50 yards. This thicket did not last for a few hundred yards or a few miles or even a score of miles. It was three-score—60 miles of iceberg puzzle.

After a full day of sledding beneath the glacier, Morton climbed an iceberg to scout a route. He could see the glacier itself from there, near its northern end. Its top was sprinkled with a great number of rocks and boulders, debris carried from the heart of Greenland. Then the two explorers resumed their determined press and, in time, found themselves in the unknown world beyond the glaciers. This region throbbed with bird life—geese and ducks filled the scene, taking advantage of every opening in the sea. More birds flew still farther north to make the most of the Arctic summer.

The explorers arrived back on the land ice, their dogs pulling them easily along, but then the ice began to melt away, until finally there was no more ice. The sea, instead of being frozen, splashed in free-moving waves along the shore. With all those birds and Hans's hunting skills, the pair might have pressed on to the Pole, but a cliff wall advanced toward the shore and blocked their way. Morton scaled one set of rocks and managed to gain an excellent view of the channel's continuation. He could see it, he reckoned, for 40 miles; and in all of that distance, he could make out not "a speck of ice." It was another misty opening trending northward, giving promise of everlasting water, and so it kept alive the world's hope for an Open Polar Sea.

Kane was excited when he heard this news of clear water, yet not quite as excited as he might have been. He had another worry on his mind. However little ice there was to the north, there was still plenty to the south. The sea was not opening up fast enough. As early as June 9, Kane had noted in his journal that the snow and ice were melting too slowly, but this observation was too terrible to consider. A month later, by July 8, he could ignore it no longer. He spent much of that day watching water dripping from a glacier. He even thought he would like to write a book about glaciers, but he knew what this too-slow melting meant. He wrote, "It looks as though winter must catch us before we can get halfway through the pack."

A second winter up there was impossible. Murderous as the first had been, the crew had had coal for heat and food to eat. Now the coal was gone, and the provisions were low. A second winter would be much more deadly.

Kane grasped at one more straw. He organized an escape party of five for an attempt to reach Beechy Island, in Canada, where they might find rescuers in the form of British ships that had come to provision that summer's northern exploration. A trek to Beechy would not be easy. The North Pole was closer. An escape to Beechy Island was the kind of impossible gamble that becomes attractive only when anything is preferable to the status quo. So nobody protested that the ships might not be there, that they might not want to come, that they might not be able to come. To raise those arguments would have been to advocate sitting still forever.

The escape party dragged one of the brig's longboats to a break in the ice and then rowed south. Kane got beyond the gates of Smith Sound, but on July 31, the group halted. A year earlier this water had been open sea. Now it was frozen solid. In the distance the men could see an enormous iceberg. Kane and McGary hiked to the ice mountain and scaled the equivalent of twelve stories. McGary had become the man Kane most trusted. He worked hard and reliably. If Kane wanted to climb a skyscraper iceberg, McGary would go with him. From the berg's top, they could look across 30 miles of frozen sea. Nothing moved anywhere in those 30 miles. There were no leads, no openings of any kind in the ice. With no way forward, the what-ifs of Beechy Island would never be tested. The whole party was trapped north of Melville Bay for another winter.

PORRENTRUY, SWITZERLAND,
SEPTEMBER 5, 1838...

Agassiz arrived at a small town in the northwestern Jura region, close to Switzerland's French border, for a meeting of the *Société Géologique de France* (the French Geological Society). Charpentier was there too. Both he and Agassiz were determined to keep the glacier issue alive. Agassiz seemed as enthusiastic as ever, preaching with the fervor of St. Paul this new doctrine of a great ice sheet that had once covered the region from the Franco-Swiss border to the plains of Italy.

How far north did the sheet extend? To Zürich? Austria? Agassiz did not say. In his Ice Age theory the sheet extended north to the top of Scandinavia, or perhaps to the North Pole. But for all his seemingly constant chugging forward, Agassiz had lost some of his nerve. He had decided, consciously and deliberately, to downplay his theory of an Ice Age. Later he would rationalize this silence as a strategy to first establish the Swiss glacier before promoting the larger concept of an Ice Age, but a letter written just before the Porrentruy meeting gave a different reason. Agassiz wrote to the meeting's chairman, Jules Thurmann, "I have decided to speak only of facts, which is too bad for those who cannot understand, unless the audience decides to discuss speculations without refuting facts which could be observed in a couple of days' traveling distance. I have too much to complain about the manner in which conscientious observations have been treated to participate a second time in such a scandal."

Agassiz was no Thomas Huxley, a man who enjoyed shoving scandalous science into the teeth of traditionalists. Agassiz, like most men, preferred applause to controversy. That was much of the reason he enjoyed life in Neuchâtel. There he could do his work and be admired by the best men in town. During his Neuchâtel Discourse he had faced up to the challenges and answered rude questions, but he never wanted to do that again. At Porrentruy he limited himself to discussing the Swiss ice sheet, providing a vivid account of his explorations earlier that summer in the Bernese Alps and on Mont Blanc. Agassiz had spent the months prior to the meeting examining Alpine glaciers, just as he had in 1836 and just as he would for many summers to come, and he brought scratched and polished rocks to show his audience in Porrentruy. He described again what had become the standard evidence—polished and grooved rocks, old moraines, and erratic boulders, all standing well beyond the current range of glaciers. Near the Grimsel Pass in the Bernese Alps, he had seen a large rock the surface of which had been polished "like the most beautiful marble."

At Chamonix Agassiz had again examined moraines that lay some distance from any glaciers. Particularly notable was a moraine a mile and a half from the front of the Glacier des Bois, in Mont Blanc's Chamonix valley. Saussure had remarked on it many years earlier, recognized it as a moraine, and marveled that it could stand so far from any present glacier.

Agassiz believed that the ancient ice sheet had moved in directions unlike those of the surviving glaciers. Today's glaciers at Chamonix flow westward, but ancient grooves indicated that when the glaciers had been larger they had moved in a southerly direction. At the Grimsel Agassiz also concluded that a glacier had crossed the Grimsel Pass, although none do today.

When the members were done reading their papers, Agassiz led a field trip along the southeastern slope of the Jura, hiking from Solothurn to Bienne, a distance of about

20 miles. The valley walls there show markings similar to the ones Kane would later sketch near Upernavik. This field trip went the way the one that Agassiz had led a year earlier, after his Neuchâtel Discourse, should have gone. Instead of brooding on the hike, Agassiz talked and pointed, showing *laves* high up the Jura slopes, indicating erratic boulders, and explaining why water could not have caused these phenomena. By the end of the hike most of the participants were persuaded that Agassiz and Charpentier were right. A glacier of fantastic proportions had once filled the region.

This success did not mean the geologists accepted the Ice Age theory in which the northern world sat under a glacier. Agassiz had argued only for Charpentier's idea, and many of Agassiz's converts would refer to the ancient glacier as "Charpentier's glacier." Charpentier himself did not believe in Agassiz's Ice Age. He rejected the theory, in part, because he thought absurd Agassiz's notion that the Alps had appeared only after the ice sheet. Charpentier also thought that the great glacier had caused Switzerland's temperature to drop, and he rejected Agassiz's contrary proposition that a drop in temperature led the glacier to grow. Finally, Agassiz's Ice-Age theory demanded that the Swiss lakes, such as Lake Geneva, had been completely frozen over. But Lake Geneva never freezes over now, except for some fringe ice along the shore, and Charpentier did not believe Switzerland had ever been so cold that Lake Geneva had frozen right down to its floor. Nor did Charpentier believe that if the lake had frozen, such an intense cold would have allowed enough water, either as rain or snow, to create and sustain so huge a glacier.

Charpentier's last two objections to the Ice Age theory are a warning against imagining his own theory as proposing a kind of local, or Swiss, Ice Age. Today's readers naturally want to assume that Charpentier thought pretty much as we do, but he did not propose that Switzerland had experienced a period of Arctic cold. He did not believe that great cold was necessary to create a great glacier. Nor did he believe that a huge ice sheet, during a period of great cold, was physically sustainable. The kind of endlessly flowing ice that pours from the Humboldt glacier or into Melville Bay seemed to Charpentier something meteorologically impossible.

At Porrentruy Agassiz had either gone on a tangent or was engaged in a flanking maneuver. His research in the Alps was establishing the science of glaciers, but it was not establishing an Ice Age. Meanwhile, however, Agassiz was becoming an expert on glaciers and their effects. He devoted his formidable energy to tramping all through the Alps. In 1839, he spent more time in the Bernese Alps, examining a great moraine at Kandersteg, and then he crossed over to the Pennine Alps to study the amphitheater where eight glaciers flow together to form the Zermatt glacier. In 1840, he climbed the Jungfrau and established a camp on the Aare glacier.

He became well acquainted with the Aletsch glacier, the largest and longest glacier in the Alps. Its runoff doubles the size of the Rhône River. Besides its familiar moraines and rock scratchings, the Aletsch has created a small lake—the Marjeelensee—whose waters were dammed up by the glacier. It showed that besides serving as a water source, glaciers could stop water. Agassiz was always picking up complicating facts like that to add to his story, confident that one day they would prove useful.

Agassiz's wife seldom accompanied him to any of these meetings, but he did not travel alone. He had a team of trained geologists and assistants who gathered data and drew their own conclusions about observations. Agassiz was pioneering something new, bringing teams of scientists together to do fieldwork. At Porrentruy it had not just been Agassiz and Charpentier who delivered papers on glaciers. Agassiz's team had been there too. One of his geologists, Arnold Guyot, for example, read a paper about the structure of glacier ice, noting a kind of banding feature in which blue and clear ice alternated throughout the glacier. So what? Nobody knew or even wondered. It was one more detail in the endless stream of facts gushing from the Agassiz team.

FREIBURG IM BREISGAU,
GERMANY, SEPTEMBER 1838...

Almost immediately after the Porrentruy meeting, Agassiz traveled to Freiburg im Breisgau for another convocation, this time of the Association of German Naturalists. At Agassiz's invitation, Lyell's old teacher, William Buckland, had also come, along with his wife. Buckland had some reason to honor Agassiz's invitations, for Agassiz had provided Buckland with extensive information about fossil fish to include in his Bridgewater book. Then Agassiz had translated Buckland's book into German and had published it through his own press in Neuchâtel, although he had omitted some of Buckland's theology from this edition.

At Freiburg Agassiz hoped to give a repeat of the Porrentruy performance for a German audience and, with Buckland present, perhaps for an English one too. Buckland was still England's best-known geologist. His Bridgewater work had well outsold Lyell's *Principles.* The first printing, at least 5,000 copies, had sold out before appearing in any bookstalls. A second printing also sold out, and a further edition had to be published. Agassiz wanted very much to win the esteemed Buckland over to the glacier theory.

And Buckland was interested because he realized that Agassiz's Ice Age idea offered a solution to the problem of the diluvium. During the 1820s, Buckland had been especially diligent in seeking out evidences of Noah's flood—stray features that might have been swept away by the flood, for example. He found many evidences, especially in the north of England and in Scotland. After Buckland concluded that those features were not the remains of the biblical flood, he was left to wonder what they were. In his Bridgewater treatise Buckland had written that

> *one of the last great physical events that have affected the surface of our globe was a violent inundation which overwhelmed a great part of the northern hemisphere....And this event was followed by the sudden disappearance of a large number of the species of quadrupeds which had inhabited these regions in the period immediately preceding it....The event in question was the last of the many geological revolutions that have been produced by violent eruptions of water, rather than by the comparatively tranquil inundation described in the Inspired Narrative.*

That was probably the first time most readers had ever seen the flood in the Noah's ark story described as "tranquil." Buckland's use of the term here hints at the many problems of the inundation that he recognized. From where had the waters come? Where had they then gone? How had they produced their changes? Agassiz's Ice Age theory promised to make sense of the diluvium by getting rid of the water. So Buckland was interested, but, as usual, words were not enough to convert someone to the idea of a great Swiss glacier. Buckland did agree, however, to come to Neuchâtel with Agassiz for a look at the Jura proofs.

Besides Agassiz and the Bucklands, a fourth traveler joined the group as it rode to Neuchâtel. He was one of Agassiz's patrons, the prince of Canino, Charles-Lucien Bonaparte, Napoleon's brother. Since the Battle of Waterloo, the prince had devoted his energies and money to science. It is typical of Agassiz's ability to make a favorable impression that he could have as patrons such ferocious enemies as the king of Prussia and the family Bonaparte. What Buckland thought of traveling with the brother of England's longtime enemy, "the thief of Europe," is not recorded. Pressed together in a coach, the foursome traveled through the Rhine valley to Basel, then across the Jura to Bienne and on to Neuchâtel. From the coach's windows they could see the occasional erratic boulders and polished mountainsides that Agassiz relied on to make his case.

Agassiz then took Buckland and the prince into the Jura for a foot tour of the glacier evidence. Despite the landmarks, however, Buckland hesitated. He did not know enough about glacier effects to say for sure that Agassiz's evidence came from a glacier. Agassiz suggested that Buckland take a look at some real glaciers. Two of the most readily accessible were up in the Aare valley, opposite Neuchâtel, at Grindelwald and Rosenlaui. Buckland set off with his wife, and when he returned to Neuchâtel, he announced that he had become a believer. The glaciers had shown him that the Jura phenomena were indeed traces of an ancient glacier. Buckland also said that these effects reminded him of similar landscapes in Scotland that previously he had attributed to floods. Although there were no modern glaciers in Scotland, he had seen rocks with glacier grooves near Edinburgh, and near Ben Nevis he knew of something that he now thought could be a moraine.

Agassiz was immediately interested. His Ice Age theory would be much stronger if he could show that Scotland had once been under a glacier, but Agassiz could not remain happy about Buckland for long. He received a thank-you letter from Mrs. Buckland that expressed appreciation for his hospitality. She added, "But Dr. Buckland is as far as ever from agreeing with you." What had happened? Second thoughts? Had Buckland begun to wonder about the theological side of the Ice Age? The letter did not say. Agassiz, however, had heard of a coup in the offing—Scotland—and he did not forget.

RENSSELAER HARBOR, AUGUST 28, 1854...
At noon a group of "deserters" (as Kane styled them) left the party to try for Upernavik before winter arrived. Most observers would use a less harsh characterization of this second escape party, as it acted only after Kane gave his permission. Among whalers it was routine to allow crewmen in a trapped ship to try to escape over the ice if they wanted to risk it.

Kane raged over the departure as a personal betrayal. The crisis of a second winter was particularly unsuited to his leadership failings. Perhaps romantics never make great leaders. They do not understand what motivates others and do not recognize how their own attachment to abstractions like loyalty or honor excites suspicion and contempt. Worse, romantics, and idealists in general, put the mission first and the people on the mission a distant second. Thus, even though the end of summer horrified him—"I cannot disguise it from myself that we are wretchedly prepared for another winter on board. We are a set of scurvy-riddled, broken down men," he confided in his journal—Kane was dumbfounded when most of the crew said they wanted to make a try to get below Melville Bay. Yet Kane was no Captain Bligh, and he let them go, giving them a just share of the provisions and letting them take one of the longboats.

The argument for leaving was simply that winter's approach was unbearable. On August 15, birds began flying south again, and who could not want to fly with them? The ice, which had never properly melted, had begun to thicken. It was maddening for men of action to sit still while the sun dripped away like water through fingers.

The case against going was that the trek was doomed. If Beechy Island had been unreachable a month earlier, why would anyone think success might be possible so much later in the season? In the best of circumstances, crossing Melville Bay by longboat might be impossible. Who could dream of crossing it once the sun slipped south of the equator?

Knowing all this, nine men left while eight stayed with Kane.

The escape group's idea was to drag a boat across the ice to open water, sail down the coast, cross Melville Bay, and continue on to Upernavik. The ice they crossed on this journey was not the frozen-pond kind of skating rinks that is familiar to inhabitants of the temperate zone. This ice moved with the tide. Isaac Hayes reported that as they crossed it, "The whole pack was grinding, squeezing, and closing." When the tides changed, the ice shrugged and moved. Leads suddenly broke through its body, turning solid footings into risky ones.

After nine days the escape party reached open water and boarded their boat, originally named the *Forlorn Hope* but renamed *Good Hope.* They passed Cape Alexander and put Smith Sound to their backs. Glaciers lined much of the Greenland coast now.

They sailed like the ancient Greeks, hugging the coast and coming ashore to cook their food. The water there displayed ice phenomena not much known in the warmer world. Hayes dubbed one kind of miniature ice floe "pancake ice" because it had the shape and size of stale flapjacks.

On September 27, a month after leaving Kane, the party members reached their southernmost point. They had gone 200 miles, and they had reached the lower end of Cape Parry, a small peninsula jutting out into Baffin Bay. To continue farther, they would have to return to rowing through open water, but the sea was too icy and the wind too ferocious. They improvised a hut and stayed where they were. Remarkably, they had some companionship. A small group of Eskimos was settled nearby, and its headman, named Kalutunah, was willing to deal with the newcomers. He was surprised by the group's ambition to reach Upernavik. He had heard of the place before and knew something of its wealth—wood, iron, and a constant supply of food to hunt. Of course, everyone would like to go there, he said. He would like to take all his people to settle permanently, but it was impossible. Between here and there was the great frozen sea—the glacial wilderness of Melville Bay. Might as well try for the moon as try to cross that uncrossable barrier.

LONDON, APRIL 1840...

Two authors were putting the closing touches on their latest books. In Neuchâtel Agassiz was finishing his *Études sur les glaciers,* the first full-length study of glaciers ever published. In London Lyell was revising his *Principles of Geology* for yet another edition, its sixth. Lyell's book had been in print for ten years now and had established itself as the leading English-language geology text. Students, teachers, and the intellectually curious had made it their standard reference on the facts of geology.

Lyell's practice of revising his old books marked one of the serious differences of approach between him and Agassiz. Like most authors, Agassiz wrote his books and moved on to new ones. He published his series on fossil fish without returning to the old volumes. His new book on glaciers would be a onetime work as well. Later, he would write again about glaciers, but that book too would be a new book, not just a revision of his *Études* with further material and corrections. But Lyell was doing something new, something in keeping with his political role of building as great a constituency as possible for his argument that all geological phenomena were the result of modern causes at their modern strengths. Facts were important in Lyell, but they were secondary to the point of promoting his viewpoint. With each new edition, he would revise and delete old material while adding new information. The effect was twofold. It proved how fully Lyell kept abreast of the latest information, and it created a strong impression that Lyell's approach to geology was withstanding the test of time and of new discoveries. On the one hand, Lyell looked like the very model of a modern scientist, altering his opinions as new facts demanded. On the other hand, he seemed the perfect old prophet as well, for all of these changes and revisions continued to support the book's overall thesis.

Over time Lyell's efforts had the kind of effect that a politician with a clear agenda can have when the opposition is not so consistent and dedicated. Because it is reasonable, the opposition makes concessions and adaptations to the maverick's telling points, but the maverick does not return the courtesy. Political historians then find themselves trying to explain why the political climate has changed without their being able to point to an incident or series of incidents that forced the shift. The change reflects years of patient effort by a dedicated few.

Science historians have the same trouble in explaining Lyell's success. There was never an observation or experiment or series of observations that defeated Lyell's opponents and raised his theory on high. But there was also never a day when Lyell was not alert to the needs and opportunities of his agenda, so over time he just seemed to grow stronger, holding more and more ground, while the ideas of his less single-minded opponents became confined to limited and special issues.

Perhaps no aspect of geology shows Lyell's success better than the role of glaciers. He had said nothing of their importance in his first edition, but as he drafted the preface for his sixth edition he wrote, "A chapter has been introduced for the first time, on the power of river-ice, glaciers, and ice-bergs, to transport solid matter and to polish and furrow the surface of rocks. The facts and illustrations contained in this chapter have been almost entirely derived from my private correspondence during the last four years or from new publications."

That private correspondence had been with Agassiz, now being turned against him in Lyell's anticatastrophe work. Lyell, of course, was never so foolish or discourteous as to criticize Agassiz directly. He simply presented the facts, organized according to his own interpretation of them. Here is how he handled the matter of erratic material: "It is well known to all geologists that many enormous masses of gneiss and other rocks are scattered over the lowlands bordering the Baltic"—he is referring, as in his letter to Darwin, to the erratics of northern Europe, and he passes over in silence the erratics of the Jura, the blocks of Monthey, and other evidence from Switzerland—"which may have been laid dry by the slow upheaval and desiccation of the bottom of the sea, which is now ascertained to be going on in Sweden and the Gulf of Bothnia."

The expert reader, who may have felt a twinge at the absence of any reference to the Jura, can now come to full attention. Lyell is adding something new to his drift theory. It does not just hold that icebergs carried the rocks. It adds the detail that the boulders had been dropped into the sea by melting icebergs, and then the sea bottom rose to become dry land. In his first edition Lyell had scorned von Buch's evidence that the land of northern Europe was rising, but a later tour of Sweden persuaded him otherwise. Now he had reversed his position on the facts, yet somehow his larger point was merely bolstered by the change. In hindsight this matter of the rising land seems especially coincidental. Later geologists would explain the land's upward movement as a continuing response to the removal of the crushing weight of the enormous ice that pressed on the land thousands of years ago. "This upward movement may, in the course of ages, have worked so entire a change in the physical geography of these countries, that it is scarcely possible to speculate on the course which drift ice may formerly have taken in northern Europe, unless we are guided by geological data. In those parts of Canada and Labrador"—now he adds North America to his analysis, while still ignoring the Jura—"where boulders are most frequent on the seabeach, being seen in great abundance at the stumps of trees wherever the forests have been partially cleared away. This phenomenon need not be attributed, as some have proposed, to the passage of a flood over the land" (by mentioning floods and not even bothering to dismiss the Ice Age theory, Lyell keeps the reader's eye off that particular ball), "but rather to a change of level, like that observed in Scandinavia, which is slowly converting the bed of the ancient ocean, long the receptacle of icebergs into a part of the American continent."

So, the "well-known" enormous masses of gneiss and other rocks scattered over the Baltic became one more bit of Lyell's general argument that what goes on now is sufficient to explain what went on in the past.

Yet this new edition also showed that Lyell still did not understand the essential difference between a glacier and an ordinary snowbank. In the book's glossary he defined a glacier as "vast accumulations of ice and hardened snow in the Alps and other lofty mountains." Even if we ignore the fact that glaciers also occur in Greenland's lowlands, we are still struck by the way Lyell missed the point that glaciers *move.*

They are not just white swatches sitting passively high up in the picturesque peaks. They are active forces crawling back and forth across the land, and that motion makes them fundamentally unlike the frozen fields of snow familiar to every English reader. The sort of throbbing, bellowing icescape that plagued Kane's party was still beyond Lyell's reckoning.

GLASGOW, SCOTLAND, SEPTEMBER 1840...
Immediately after publishing his *Études*, Agassiz traveled to Britain for the annual meeting of the British Association for the Advancement of Science, held that year in Glasgow. As usual at such conventions, the socializing and private back-and-forths were at least as important as the formal papers. One young naturalist, Edward Forbes, boasted in a letter after the meeting that he had spent the whole week in Glasgow chatting with Agassiz on the technicalities of sea-life classification. Agassiz's reputation in England still rested on his fossil fish, not glaciers.

On September 21, Agassiz presented four papers to the Glasgow meeting. He read one of them before the geology section, where he spoke in French about the glaciers and boulders of Switzerland. Lyell presided, so, of course, this time there were no boos or catcalls. Yet the theory was still scorned. If Agassiz's abundance of new facts did anything, it provided another target for his audience to challenge, making persuasion even more difficult. His explanation of glacial motion, for example, did not satisfy everyone in the room, especially those with a solid grounding in physics and dynamics. Agassiz was no physicist and was set slightly on the defensive when questioned on the matter, but the important fact for the Ice Age theory was simply that glaciers moved. How they moved hardly mattered. Scientists, of course, do not like to hear that new facts are irrelevant to their understanding, but Agassiz had presented all the major ones at Neuchâtel before he ever went into the field to do his own glaciological research. The critical fact supporting prehistoric great ice was that glaciers move; and as they move, they (a) push moraines in front of them, (b) carry boulders on top of them, and (c) scratch and polish rocks beneath them. The mechanics of glacial motion, transportation, and scratching were technical details that could not be decisive in proving or disproving Agassiz's theory of an Ice Age.

The question before the Glasgow meeting was the same one raised at Neuchâtel: Had the effects Agassiz cited been caused by glaciers or by something else? Moraines were the least controversial souvenir of glaciers past, and Lyell proclaimed himself satisfied that Agassiz had provided a complete explanation for how they had been formed. Yet in the same breath, he still doubted that even Switzerland's glaciers had ever been as large as Agassiz supposed. Lyell suggested that small local glaciers could account for the moraines that now stood far away from the existing glaciers. Discouraging as this dismissal was, it would prove Agassiz's best news of the day. None of the other evidence of glacier action won even this limited acceptance.

Erratic boulders still were not taken as even suggestive evidence. Partly, this resistance came from a professional skepticism, but much of it reflected ignorance of what the tops of glaciers looked like, and still more of it erupted from a commitment to other explanations. Roderick Murchison was the most vocal opponent of Agassiz's account. He was defending his own use of drift theory, and part of his attack was gamesmanship. After the meeting he wrote a letter saying, "Agassiz gave us a great field-day on Glaciers, and I think we shall end in having a compromise between himself and us of the floating icebergs." Agassiz did not realize that Murchison was setting up a negotiating position; he assumed that, as Cuvier had preached, he could carry the day with the facts of the case.

As for the evidence from polished rocks, the doubters did not even see them as a source of puzzlement. Lyell objected that such stones were just too plentiful to be significant. He himself had seen many such rocks in Scotland and Sweden, yet Scotland had no glaciers, and Sweden's polished rocks, seen in fjords, might have been caused by icebergs trapped and then twisted against the rocky coast.

William Buckland sat quietly through the session, offering Agassiz no help. He had done nothing to explore the glacial idea since his tentative conversion two years earlier, but Agassiz had not come to Scotland just to be scorned one more time. He finished his presentation by announcing that he planned to tour the Scottish Highlands and was confident of finding many evidences of a past glacier. Buckland was going to accompany him.

The two geologists became partners and moved into the chilly, desolate Highlands as soon as the Glasgow meeting ended. They were seeking proof of stage two of the glacial theory—that there had once been glaciers where none now existed, even in diminished form. Their route took them north from Glasgow to the Great Glen. Then they followed Scotland's major fault line right up from Fort William in the south to the Moray Firth in the north. For a Swiss, the

Scottish landscape was second-rate, with even Scotland's most beautiful regions, like Loch Lomond, receiving their strongest praise in the observation "Reminds Me of Switzerland"—Switzerland without the Alps, that is. But for a pair of geologists looking for evidence of glaciers past, it was a paradise. The Scottish Highlands, in every glen and pass, bear witness to an Ice Age. For these explorers, the absence of an Alpine range was to the good. Scotland's Highlands never reach an altitude of even 4,500 feet, so where had all the ancient ice come from?

In a letter to Humboldt, Agassiz reported that Scotland was just like Switzerland: "I found the same polished surfaces as in Switzerland, the same lateral and terminal moraines spreading in the same spoked array coming from the center of mountain chains toward the plain, and, throughout, these lakes too show a reduced sedimentation because of the glaciers that once occupied their depths."

The partners went by Loch Lomond and across the Rest-and-Be-Thankful Pass that leads from Loch Long to Loch Fyne. Agassiz may have suspected the Ice Age origins of these lochs, but he was cautious. Only as they approached Inverary Castle, home of the Dukes of Argyll, did he say to Buckland, "Here we will find our first traces of glaciers." Sure enough, the carriage road led across a classic terminal moraine. Both partners recognized it for what it was, and both knew what its presence meant for the Ice Age theory.

A bit farther north they reached Ben Nevis, Great Britain's highest mountain. It is only about 4,400 feet tall, but at that latitude the height is enough to assure a constant snowcap, although not a glacier. Still, the partners could see that glaciers had once run down all the mountainsides, for moraines, erratics, and polished rocks spread out from Ben Nevis like rays from the sun, and Ben Nevis is still just in the southern end of the Highlands. It serves as a gateway to the glens and lochs above. Yet even in Scotland's south, the two geologists had already seen enough to know that the whole of Scotland's geological history was going to have to be reinterpreted in terms of an Ice Age.

Sixteen years earlier Buckland had traveled through this area with Lyell and had thought that he had seen and understood much, but the glacier theory gave new meaning to the land's appearance. As was typical of Buckland's open-minded approach, he seems not to have fought the glacier explanations but calmly watched his old ideas and explanations disappear the way Scotland itself had once vanished beneath an icy wall.

By now the partners needed only some pretty little cherry to set on top of their geological revolution, a flourish that would add something unexpected to the wealth of moraines, boulders, and scratch evidence they had amassed. They found it when they came to the parallel roads of Glen Roy. These "roads" are perfectly straight lines that run across the valley walls. Each horizontal line is matched by another on the opposite slope. Their straightness and shared altitude seem more like something for a geometrically gifted engineer than nature. Lyell, who had seen the glen during his journey with Buckland, had written his mother that the roads were "one of the grandest natural phenomena in Great Britain." The local people attributed them to some lost tribe of ancient builders, but most geologists thought they were lake terraces. They just could not figure out how the glen could ever have been filled with water.

Agassiz recognized the roads as the spectacular remains of a great glacier dam, like the Marjeelensee that he had seen in Switzerland. He told Buckland that a glacier, probably extending from Ben Nevis, had dammed up the glen, forcing water to rise until it reached high enough to spill over the sides of the ridges. He proposed that the "roads" marked the various stable levels that had once been reached by the lake. Buckland had heard scores of attempted explanations for the parallel roads. Charles Darwin had recently published one himself. Usually these explanations included some kind of inundation or flood. In Darwin's case the explanation was that Scotland had sunk and that seawater had rushed in to lap against Glen Roy's shores. Agassiz's explanation was wholly new, unexpected, and yet in keeping with the revolution the two of them had been preparing. On Buckland's encouragement, Agassiz sent a letter to Edinburgh reporting his explanation for the parallel roads, in those days one of the most famous natural sights in Scotland. His account appeared in a newspaper, *The Scotsman,* four days after being posted, and it began causing a commotion even while the two explorers were still deep in the Highlands.

Meanwhile, Agassiz and Buckland had made their way up the Great Glen, beyond Loch Ness to the Moray Firth on Scotland's northern coast. This wilderness held a great estate called Altyre House, the home of an enthusiastic naturalist with a fine collection of fossil fish, so naturally Agassiz stopped by to have a look. The partners arrived at Altyre House full of enthusiasm for their observations of glacier remains, and when they discovered that Murchison had come there from the Glasgow meeting, they bubbled over.

Moraines, erratics, scratched rocks, and the parallel roads of Glen Roy were set before Murchison's imagination, but he would have none of it. Gamesman to the end, he held firm. No. Never. Cannot be.

For any man who was less of a steam engine than Agassiz, Murchison's rebuff would have been painfully sobering, but Agassiz was uncooled. He turned his attention to another guest in the wilderness mansion, a young physicist named James Forbes, the same Forbes who had met Ignace Venetz while hiking in the Alps a decade earlier. Forbes too had been at the Glasgow meeting and was considered one of the bright young men of British science. Although he was no geologist, Forbes's knowledge of Switzerland and his familiarity with glaciers made him interested in Agassiz's theory. As a physicist he took it for granted that glacial motion must be having some effect on the land beneath it, although, even after his conversation with Venetz, he had never paid any notice to glacier effects in Switzerland. For Agassiz, a wink was always as good as a nod, and he urged Forbes to come to Switzerland to have a look for himself. History remembers this offer because it was to have consequences, but Agassiz was probably not being especially perceptive when he made it. Instead, it seems more accurate to picture Agassiz as a missionary for the Ice Age, a man who responded to every small show of interest with an invitation to embrace the new knowledge.

The partners then left Altyre House and made their way south. It did not take long for Buckland and Agassiz to see further evidence that they were right and Murchison wrong. They found still more moraines. Eventually the two men separated. Agassiz took a ferry to Ireland, ostensibly to examine yet another collection of fossil fish, though he was soon writing Buckland with reports of glacier evidence in Ireland too.

Meanwhile, Buckland made his way to Lyell's home in Forfar, in southern Scotland. The estate, Kinnordy House, was Lyell's childhood home, and he loved it with the confidence of a husband who believes he knows every detail of his wife's nature. Buckland promptly showed him better, by pointing out a long, slender, low ridge that Lyell acknowledged had always been a mystery for him. It was, Buckland explained, a moraine, the remains of a glacier standing only 2 miles from Lyell's beloved home. Lyell saw it was true; at some time in the past, a glacier had stood right there.

Lyell's abrupt conversion was more an act of recognition than a feat of reasoning. All of these early believers in the glacier theory had been changed by this double process of looking and then seeing. Charpentier, Agassiz himself, Buckland, and Lyell had been doubters and mockers. Then suddenly, like the secret image hidden in a child's puzzle, the truth popped out at them, and they could recognize what had been before their eyes all along. *I once was blind but now I see.* After that change, their perception became as impossible to deny as the roar of the ocean or the brightness of the sun. Glacier remains became simply one of the givens of the landscape.

For Buckland, Lyell's change was a prayer answered, and he immediately wrote to Agassiz that "Lyell has accepted your theory *in toto.*" With that news singing in his heart, Agassiz returned to Scotland and traveled to Edinburgh, where he was pleased to be greeted at last as the champion of a great idea.

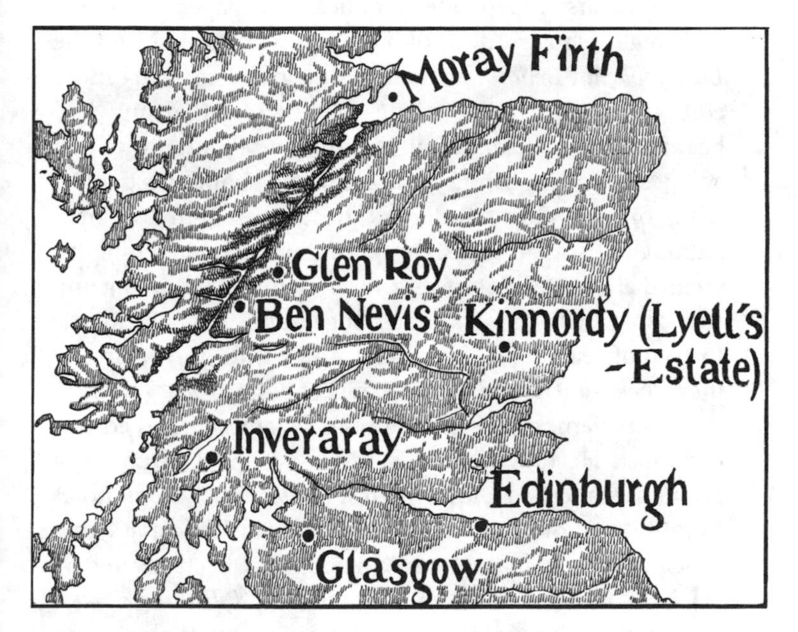
Moray Firth
Glen Roy
Ben Nevis
Kinnordy (Lyell's
-Estate)
Inveraray
Edinburgh
Glasgow

Edinburgh

Agassiz's arrival in Edinburgh has the flavor of some outland genius finally stumbling into Greenwich Village or the Seine's Left Bank, where there are artists who can appreciate what he has achieved. Edinburgh had its own geological society and two especially prominent naturalists. One was a newspaper man, Charles Maclaren, editor of *The Scotsman,* Scotland's first independent, radical newspaper. He was a self-taught scholar, editor of one edition (1820–1823) of the *Encyclopaedia Britannica,* and a contributor to many more editions. He had been at the Glasgow meeting and had heard Agassiz's presentation. It had been Maclaren who had accepted and published Agassiz's letter about the parallel roads of Glen Roy. Maclaren's greatest strength was that, although he was almost sixty, new ideas still excited him. He seemed always ready to toss aside a lifetime of thinking and supposing in favor of a better suggestion from elsewhere. Instead of reacting to Agassiz's Ice Age with horror or dismay, he was intrigued by this concept of great ice.

Robert Jameson was Edinburgh's other great editor-naturalist. He taught natural history at the University of Edinburgh and had a long interest in glaciers. Student notes from 1825 show that, even then, he told his classes that Scotland showed many signs of having once been covered by glaciers. He founded and edited the *Edinburgh New Philosophical Journal,* which, for over a decade, had been translating and publishing articles about glaciers. Jameson had printed an English text of Agassiz's Neuchâtel Discourse, not as a report of a scandal, but as something worth considering.

These Edinburgh naturalists were delighted to escort Agassiz through their suburbs and have him confirm long-held suspicions. Moraines, erratics, and telltale scratches surrounded Scotland's capital.

The glacier theory dominated the next three biweekly meetings of the Geological Society of Edinburgh. At the November 4 meeting Agassiz presented a general overview of this theory and some Scottish facts to support it. Buckland then read the first of three parts of a long paper reporting the many observations he and Agassiz had made. These meetings seemed so important that Agassiz stayed on in Scotland. Lyell joined Buckland at the second and third meetings in reading a paper about glacier traces in his home shire.

Thus, for a month, England's two best-known geologists were in Edinburgh making the case for Scottish glaciers. Lyell said that Scotland must have looked like Antarctica, with only its tallest peaks rising above an ice sheet of fantastic thickness. It was a bold analogy since the Antarctic was barely known in 1840. Agassiz was even bolder, comparing Scotland to Greenland. He imagined the Greenland interior as a great ice sheet oozing in all directions, and he suggested that Scotland had once been the same. These remote, obscure images of great ice suggest that both naturalists had realized that a Mer de Glace grown yet larger was not enough to account for the kinds of landscapes they saw in Scotland. Apparently they had managed to see in their mind's eye what Kane would observe more than a dozen years later, but, of course, the other naturalists in the room still heard "glacier" and thought "Switzerland." They had no notion of what either Agassiz or Lyell was talking about.

After the second meeting Agassiz could smell victory so sharply that his nostrils pulsed. He wrote a letter saying that the objections raised by Murchison and others at the meetings were too far-fetched to have much effect. Buckland, however, had been less sanguine about their reception and lost his temper in frustration during the second meeting. Seeing the incredulity that greeted his evidence of scratched rocks, his usually cheerful temperament gave way, and he hissed that anybody who doubted that ice had scratched, grooved, and polished the Scottish rocks should be damned to the pains of eternal itching without the privilege of scratching.

For the fact was that, despite the prestige of the theory's defenders and the mountain of evidence they offered, the meetings were going badly. As always, it is a little hard for moderns to understand exactly why. Yes, some geologists asked some good, probing questions, but why didn't the sheer mass of facts and the glacier theory's explanatory power have more impact? Between 1840 and today, society has passed through a kind of perceptual gate, and worlds that we can now sense in a glance were enigmas then. Agassiz, Buckland, Lyell, and the other glacialists were forgetting that they themselves had been converted by acts of recognition rather than by reasoned argument. They thought they needed more and more facts to tell their audience what they knew, when what they really needed was the language of a poet that could make their audience see what they had seen.

Charles Maclaren was one of the few who understood what was needed, and at the end of 1840, he published a series of reports on how glacialism offered a new way of seeing the Scottish landscape. "Even though M. Agassiz's opinions should not be fully established," Maclaren wrote, "they still afford us a new geological front of great power and widely applicable, which may help us to an explanation of some phenomena very difficult to account for with our existing means of information."

It was not enough. In February 1841, the Geological Society of Edinburgh voted a declaration that "the Glacial Theory of Agassiz is not applicable to Scotland." Even more debilitating was the loss of heart by the theory's two greatest converts. Buckland had planned a book on the Ice Age, but the wind left his sails. The idea was too controversial, too liable to evoke strong emotions, and he fell into a permanent silence about Ice Ages. Lyell too lost his nerve and retreated from his very public embrace of Agassiz's theory.

By the spring of 1841, it was apparent that Agassiz's great Scottish exploration had come to little. For the next several years, British geologists did feel a compunction to include a paragraph explaining why the glacier heresy could not solve this or that question, but these rejections were offhand and reflected no serious thinking about what the role of glaciers might have been.

RENSSELAER HARBOR,
GREENLAND, OCTOBER 21, 1854...

The end of summer's light was a clear omen of the doom that was threatening the remaining occupants of the *Advance.* "The sun has ceased to reach the vessel," Kane wrote in his journal. "The northeastern headlands or their southern faces up the fjords have still a warm yellow tint, and the pinnacles of the icebergs far out on their floe are lighted up at noonday; but all else is dark shadow." It was typical of Kane's literary skill to unite a concrete account with metaphorical language, making the last light of summer remind a reader of an approaching death.

Kane could not know that the escape party's members had already been trapped for a month in even worse shelter than he suffered, although he knew their cause had been hopeless, and he was sure that they were suffering somewhere far short of their destination. He spent no pity on them, wished them no mercy. Instead, he still raged against their treason, taking their wish to get home as a personal affront. The escape party's members became living expressions of all the failures and frustrations of this expedition. Kane vented a wrath in his journal that seems insane for its intensity, its blindness, and its repetitiousness.

Coming after so many other failures, the escape party's disappearance emphasized Kane's weaknesses as a man and as an explorer. Celebrity and Tennyson were no substitutes for the real heroism that achievement in the Arctic demanded. The smallness of the expedition's accomplishments was summed up in September when Kane posted a message relating the party's fate to any searchers who might someday come. The crewmen built a cairn on an island near the brig, and in it placed a message:

> *The labors of the expedition have delineated nine hundred and sixty miles of coastline, without developing any traces of the missing ships or the slightest information bearing upon their fate. The amount of travel to effect this exploration exceeded two thousand miles, all of which was upon foot or by the aid of dogs.*
>
> *Greenland has been traced to its northern face, [Kane took the Humboldt glacier to mark Greenland's end!] where it is*

connected with the farther north of the opposite coast by a great glacier. This coast has been charted as high as lat. 82°27' throughout its entire extent. From its northern and eastern corner, in lat. 80°, long. 66°, a channel has been discovered and followed until farther progress was checked by water free from ice. This channel trended nearly due north, and expanded into an apparently open sea, which abounded with birds and bears and marine life.

The death of the dogs during the winter threw the travel essential to the above discoveries upon the personal efforts of the officers and men. The summer finds them much broken in health and strength.

Jefferson Temple Baker and Peter Schubert died from injuries received from cold while in manly performance of their duty.

Having posted the news, Kane returned to the brig to rage, to starve, and to watch the sunlight disappear. There was little reason, beyond animal hope, to expect to be alive when the sun would return.

As darkness took over, there was less and less to do, so the men stayed still and cold, like terminal patients in a makeshift hospital. Adding to the discomforts of their bed was an abundance of rats, which were drawn like beetles to the brig's light, heat, and scraps. Rats were everywhere and more of a threat to the stores than the dogs had ever been. The men were appalled, but then Kane began trapping them and feasting on rat meat. He fed on rat drumsticks, little rat steaks, and rat hams. He urged the others to join him, insisting that nourishment was nourishment and that the refusal to eat rat was mere superstition. The others were too disgusted and literally preferred a miserable death to such wretched food. Kane had the rats to himself.

Even more deadly was the temperature. The air stayed deep below zero, and with the coal gone, there was not enough fuel to keep them alive. In early December, they began cannibalizing the brig, removing some of its planks for firewood. At first Kane told himself that there was enough extra wood in the ship to provide heat and still sail out in the next season, but, of course, that was a dream. Once they began burning pieces of their ship they were like drug users who cannot stop a bad habit, despite having begun with promises of just a taste. They knew they were eating their seed corn, but it was the only way to survive until planting time.

On December 7, a little more than three months after the escape party's departure, two of the "deserters" reappeared at the ship. They had abandoned the hut where they had been trapped since late September and had traveled through snowstorms and monstrous cold to return to Kane's shelter. Surely they would have died in the darkness, but Eskimos with dogsleds rescued them and carried them to the ship. The others in the escape party were coming back too, and within a few days all of the deserters were back, begging warmth of the brig *Advance,* which was burning down, one board at a time.

Often Kane had fantasized about just such a tail-between-the-legs return. The traitors had done with him. Let him have done with them. *Get out! Sleep in the snow and be damned!* That had been his vengeful dream, like King Lear raging upon the heath against all who had betrayed him. But now that the miserable, desperate men were back, Kane took them in at once. It was not quite the behavior of the father welcoming the returning prodigal. There was no fatted calf to eat, and Kane was careful to rank the faithful above the returnees. But he took the traitors back and shared with them the little that the party still had.

From that time on, a new, more self-critical tone began to appear in Kane's journal. He started to blame himself for the party's desperation.

> *Our own sickness I attribute to our civilized diet; had we plenty of frozen walrus I would laugh at the scurvy. And it was only because I was looking to other objects—summer researches, and explorations in the fall with the single view to escape—that I failed to secure an abundance of fresh food. Even in August I could have gathered a winter's supply of birds and cochlearia. From May to August we lived on seal, twenty-five before the middle of July, all brought in by one man: a more assiduous and better-organized hunt would have swelled the number without a limit. A few boat-parties in June would have stocked us with eider-eggs for winter use, three thousand to the trip; and the snow-drifts would have kept them fresh for the breakfast-table. I loaded my boat with ducks in three hours, as late as the middle of July, and not more than thirty-five miles from our anchorage. And even now, here are these Esquimaux, sleek and oily with their walrus-blubber, only seventy miles off. It is not a region for starvation, nor ought it be for scurvy.*

On December 25, Kane's self-accusations and reconsiderations reached their climax when he had an intense vision of his home and family. This kind of mystical experience was most unsettling to a man of action and science. He hardly dared confess even to his journal about having a vision, and he never mentioned it to anyone. "I dread the non-practical mystified atmosphere of the whole matter," he revealed in his journal. Even today this kind of experience puzzles many people, although it is among the most natural events, especially regarding desperate and dying people. It is not surprising that a starving man would, on Christmas Day, imagine in all its clarity and detail a scene of home.

On what he took to be his deathbed, Kane had a most anti-Ulyssean vision. Tennyson's poem opens with the old Ulysses back home and bored to death with domestic tranquillity, eager to go forth again. Inspired by that eagerness, Kane had made sailing beyond the sunset his own ambition, but as he lay freezing and feeding on rat meat, Kane remembered and wanted the satisfactions of all that he had fled.

Even the ship's animals were in a domestic mood. There were very few dogs left by then, and Kane used the carcasses of the dead ones to feed their living brethren. He found one of them, not quite dead, lying motionless and staring at the light that slid through a crack beneath the door. Kane bent toward her, and the dog, for the first time ever, licked his hand.

The ship by now had been cannibalized for over a month and would never carry anyone away, nor was it certain how many, if any, would live to see the daylight. When sunrise was still a month and a half away, the party began burning its lamp oil for heat as well as for light. Spring temperatures were at least three months off.

It is a wonder that the men had lasted this long. They were enduring, even by Eskimo standards, a catastrophic winter. Kane had imagined the Eskimos to be sleek and well in their villages, but his neighbors were as desperate as he was. One village had begun to kill and eat its dogs, an act even more terrible than burning the brig for firewood. When Kane did learn of what had happened and how only twenty dogs now existed in the whole of the Eskimo world between Melville Bay and the Humboldt glacier, he was overwhelmed by a great pity for these people who faced up to one of the most merciless climates on earth, and he asked about their dogs, "What can they hope for without them?" He had forgotten himself and his romantic ego completely.

Without ego or ambition, he could see now where he had landed—it was hell frozen over. Nor did it turn any holier after sunlight returned. "It is a landscape," Kane now believed, "such as Milton or Dante might imagine,—inorganic, desolate, mysterious. I have come down from the deck with the feelings of a man who has looked upon a world unfinished by the hand of its Creator."

Yet somehow in this rough draft of an earth, Kane himself had finally become a fully finished personality. He no longer saw things simply in terms of what they meant to his ambition. As he was preparing for a major hunt, Hans asked for permission to visit one of the Eskimo villages. He wanted, he said, to get a stock of walrus hides. Kane readily gave permission, even though Hans's skills would be sorely missed on the upcoming hunt. Hans refused to take any of the party's four surviving dogs (out of sixty-two with which they had arrived) and walked off into the snow. Kane never saw Hans again. He learned from other Eskimos that Hans had found a woman to his liking and had returned to family life, leaving Kane and the others to their fates.

In the previous year a desertion by the party's most valuable member would have produced a fury, but Kane had more respect now for his own limits and for the motives of the settled life. He contented himself with the remark, "Alas for Hans, the married man!"

Part IV

Rock Bottom

At this point in the history I have begun to remember just what it is about machinery that is so appealing. Mechanisms are consistent and, therefore, reliable. Computers may suffer from garbage in, garbage out, but if quality goes in, quality does come out. With people, you can never be sure, which must be why societies have always dreamt of finding heroes. With them, it is all rising to the occasion and producing the real goods. The rest of us can go along for the ride, benefiting from their reliability without having to rise to the occasion ourselves. I may be a sniveling sort, but I am safe from gestapo terror thanks to the free lunch I have been provided by the heroes of Patton's and Zhukov's tank forces. Likewise, I am no rocket scientist, but I can still enjoy photos taken on Mars, one more free meal for my mother's nonhero child. But I see now that with genius heroes, it is not quite the same. They can go out like Agassiz and Charpentier to perform their feats of recognition, grasping the meaning of what had seemed mere noise in the signal, and they can add to their own understanding where before there was emptiness, but they "heroicize," if such a word exists, for themselves alone.

The problem of progress in understanding turns out to be more profound than merely overcoming the business about garbage in. If we are to join in the new understanding, we too have to recognize what these genius heroes saw, and we too must appreciate its meaning. Looking back on his life from the vantage point of the early twentieth century, the historian and writer Henry Adams saw much of it as the constant changing of his mind. The pace of other people's genius demands constant work on our own part if we are to remain, as Adams wanted to be, educated, active participants in our own civilization. By the time he was sixty-three, the whole world Adams had been born into had disappeared. He said he "had surrendered all his favorite prejudices and...had accepted...Pteraspis and iceage and tramways and telephones." Adams hid much about the struggle to learn behind that word "surrendered."

NIAGARA FALLS, AUGUST 7, 1841...

Charles Lyell had never been one to surrender easily. His acceptance of the Scottish glacier had come on with the speed of love, and then it turned into one of those painful romances in which people ask you to your face, "What do you see in that one?" Then he fell silent. Now he was aboard a train approaching what seems today North America's most spectacular child of the Ice Age. The train offered its first view of the gorge 3 miles below Niagara Falls and sped him on to the wonder itself. The Falls in those days were louder and wilder than the tamed river that visitors meet today, when over half the flow is diverted into the pipes and sluices that supply Niagara's hydroelectric plant. Erosion at the Falls since 1841 has been especially strong along the sides, so that the Canadian falls, although still known as Horseshoe Falls, have lost much of their U shape. The visitor today, seeing Niagara's enormous flow, its perpetual rainbows, and the constant fog that rises from beneath the Falls, is impressed, but this splendor is only a souvenir of the sight that greeted Lyell. The scene won him for good. Today it is common for visitors to spend no more than fifteen minutes peeking over the gorge before going on to the shopping outlets. Lyell stayed for a week and then came back later for a second week of exploring.

By 1841, Niagara Falls had already become a popular tourist sight, although the mobs then were nothing as incredible as today's crowds, thick as a packed subway, that press against the rim. In Lyell's day the point beside the Falls was private property, and visitors had to pay the hotel keeper for the right to pass beyond its high fences and stand beside the drop.

Lyell was, in part, a typical tourist, gawking at the Falls, buying a daguerreotype of the scene to carry home, and enjoying Goat Island (the strip of land between the American Falls and Horseshoe Falls) as "the most perfect fairyland I know." In part also, he was a geologist visiting one of the world's most remarkable geological spectacles. He toured the area with James Hall, a geologist from the New York Geological Survey who was the most knowledgeable authority on the Falls then living. Together Hall and Lyell dug up fossils on Goat Island and the Canadian bank. They explored up and down the gorge. Geologically speaking, the gorge is more astonishing than the Falls themselves. For one thing, its 300-foot walls show the beautiful stratification of the rock—tier upon tier, going down to an ancient red sandstone. Even a geologist of Lyell's experience and travels had rarely seen a sharper illustration of the earth's stratification.

Seven miles below the Falls stands the Niagara escarpment, a 300-foot-high cliff wall that runs for hundreds of miles to the northwest. Well before Lyell's visit geologists had realized that the Niagara Falls had once tumbled over the escarpment itself and that, during uncounted thousands of years, the cataract had carved its way back to the Falls' present point.

Also, erratic boulders are sprinkled across the whole area like salt on french fries. Still today anyone driving a car through the little town of St. Davids, at the base of the Niagara escarpment, can see the boulders without even slowing down. It looks as though people have decorated their yards and church grounds with randomly placed large rocks.

There are also grooved and polished rocks aplenty in the area. Most visitors never stray far from the standard paths, but anyone today who hikes into the gorge by the Whirlpool Rapids can see such rocks lying just beside the trail. There is even a great moraine on the Canadian side of the river, rising steeply immediately behind the Falls. A tram carries today's visitors up its slope.

Lyell saw all these elements, of course, and knew exactly what Agassiz would have made of them: They offered the opportunity to extend the Ice Age theory to North America. But Lyell's attitude toward that theory had become uncertain. In his most recent edition of the *Principles,* he had said expressly that the North American erratics were the result of icebergs and land uplift. But he had not been to America when he wrote that, and he had described the boulders as a coastal phenomenon. Niagara Falls is not coastal. After issuing his sixth edition, Lyell had sided with Agassiz and Buckland in Edinburgh, but then he grew silent as the glacier idea was branded heresy.

There was another issue as well. If the Scottish glacier had come on Lyell with the force of love, it was a forbidden love. The Ice Age theory denied everything for which Lyell had previously stood. It was catastrophic, and it imagined glaciers of a size and power unlike anything he knew on earth. Lyell had compared Scotland to the Antarctic, but how had Scotland, so far from either Pole, ever grown so icy? On the other hand, Lyell had recognized a moraine near his family home. So what was it going to be: his eyes or his prejudices?

Lyell did not surrender to his eyes. At Niagara Falls he completely rejected Agassiz's notion of an Ice Age. He interpreted none of the Niagara phenomena, not even the moraine, as remnants of a glacier. Yes, he had once advocated the idea of a glacier covering Scotland and had seemed to have converted to Agassiz's cause, but he was an apostate now, and he returned to explaining all of Niagara's markings in terms of his old idea of water, drifting icebergs, and changing elevations. Lyell said that the whole region had "gradually submerged." Water came in with icebergs floating after, and "the country was then buried under a load of stratified and unstratified sand, gravel, and erratic blocks." Afterward the country rose again—slowly, to be sure—leaving erratic residue scattered at random, hither and yon.

Lyell did not try to explain what machinery could raise and lower so enormous a chunk of the North American continent without producing any cracks or folds in it. He did notice that the icebergs had produced astonishingly parallel grooves that, everywhere in the Niagara region, moved from northeast to southwest, but he contented himself with reports that icebergs are very faithful to their course.

Agassiz had argued that the Ice Age must have been geologically recent because its effects still lie on the surface. They have not yet been buried by newer phenomena, and the fossils associated with them are contemporary. Lyell agreed that the evidence pointed toward relative recency and was moved to remark, "If such events [the submergence and resurfacing of North America] can take place while the zoology of the earth remains almost stationary and unaltered, what ages may not be comprehended in these successive tertiary periods during which the flora and fauna of this globe have been almost entirely changed."

Lyell did make one change in his drift theory that can confuse today's readers. Instead of speaking directly of "drift theory," he wrote of a "glacial theory" and said that the submergence "took place during the glacial period, at which time the surfaces of the rock were smoothed, polished, and furrowed by glacial action." To our ears, "glacial" implies glaciers, but that interpretation goes along with our knowledge of great ice and our acceptance of the Ice Age. Historically, however, "glacial" simply means icy and cold. In fact, the *Oxford English Dictionary* dates the word "glacial" as far back as 1656, whereas "glacier" first appeared in English almost a century later, in 1744. This confusion of terms means that today's readers must stay alert when they read documents of 150 years ago. A reference to a "glacial period" or "glacial action" sounds familiar, but the writer may have been imagining something more like Maine's winter coast than Greenland's ice sheet. By "glacial action," Lyell referred to ice, to bergs, and to floes in salt water. By "glacial period," he did concede Agassiz's notion that the earth had been colder, frigid enough to allow sea ice to drift beyond its present limits, but he did not concede the presence of a great glacier.

While at Niagara, Lyell did make one great geological discovery, at the Whirlpool. The Niagara gorge behaves most oddly there. The river makes a sharp turn, entering the Whirlpool on a northwestern course and leaving on a nearly perpendicular route to the northeast. It is obvious to anyone with a geological imagination that something happened at the Whirlpool to change the Niagara River's erosion pattern. The Whirlpool defies nature. The water jerks and spins like a running dog suddenly choked by a master's leash. Meandering rivers are familiar, but this sort of zigzag seems to contradict physics.

Lyell and Hall examined the Whirlpool gorge together. On the Canadian side, an explorer's eyes are drawn to the gorge. Water from the Falls is rushing straight for this wall and creates a splendid view that is normally available only to birds. Lyell, however, was enough of a geologist to look at the ground at his feet, and he pointed out to Hall that the stratification had suddenly gone missing. Sharply defined layers of rock were visible on all sides of the Whirlpool, but the area directly facing the Niagara River's inflow was composed of a mixture of mud and erratic rocks of every size. Hall immediately guessed that this break in the stratification had something to do with a similar break in the escarpment's stratification above St. Davids. Here was an unexpected discovery worth exploring in detail. Unfortunately,

Lyell's time in Niagara was up, but he left vowing to return and to explore the mysterious breaks in stratification.

Lyell proceeded on his grand tour. He reported that erratic boulders persisted as far south as Long Island and then disappeared. The American South showed no signs of the drift. Lyell toured the whole country, making social observations as well as geological ones, but his English fondness for privilege and rank made him an indifferent interpreter of American democracy, and his commentary had none of the impact that his geological report enjoyed.

In June of 1842, Lyell returned to Niagara Falls to explore the link between St. Davids and the Whirlpool. He hiked the 3 miles between the two points and concluded that unstratified soil filled the whole space. Wells dug in this area went through a mud-and-rock mix, not through layers of stone. Lyell had discovered that there had been another Niagara gorge before the present one. Somehow that old gorge had been completely filled in with a mixture of dirt and rocks. Something had come along with the force and thoroughness of a railway gang to seal up the old gorge. Later, water dug a new gorge, one that cut its way back from the escarpment until it reached the old gorge and began to reexcavate the old river's former route. That is why the Niagara makes such an impossible turn at the Whirlpool. The two routes had been created in two distinct epochs.

The old gorge is 2 miles wide at the St. Davids notch. What kind of hand could have patched up so wide a crack? Did a glacier cross Lyell's mind? Yes, but he rejected it. Niagara and New England were "too remote from any high mountains, [so] we cannot attribute effects...to true glaciers descending in the open air from the higher regions to the plains." Glaciers, for Lyell, still meant Alpine white caps descending short distances toward the plains below. As there were no Alps anywhere around Niagara, or Boston, or Long Island, there could be no glacier debris.

It is a shame that Margaret Fox's rapping proved a fraud, for it would be enlightening to summon Lyell's spirit and ask what about the moraine he had seen near his old homestead. How could he repudiate that recognition?

A clue to the spirit's likely answer comes in Lyell's account of the St. Lawrence River valley, which was filled with erratic and scratched rock that he knew Agassiz would explain in terms of a great glacier. Lyell, however, said only, "I know of no theory that can account for [erratics associated with grooved rock], with any plausibility except...the agency of large islands of floating ice." He dismissed the glacier theory as implicitly "implausible," a two-dollar word often used in place of real argument. After all the years and facts, opposition to the idea of great ice was still as intuitive and dismissive as when James Forbes first met Ignace Venetz and rejected out of hand his theory of a Swiss glacier. Lyell had recognized the moraine, but what had he really seen? An unstratified ridge of dirt, very like a moraine; but his reason told him it could not *be* a moraine. For all his greatness and influence, Lyell's principles could never let out more than they let in.

So he explained the old Niagara gorge the way he explained all the other Niagara debris: The region had sunk; icebergs had labored like determined barges, dropping tons of dirt and rock into the old gorge. Then the elevator had come back up, the sea retreated, and a new gorge was cut.

After he left Niagara, Lyell went on to Toronto to examine raised beaches and strange narrow ridges capped by erratics. His guide believed the area had once been buried under a freshwater lake. Lyell recognized the raised beaches as matching, on a giant scale, the parallel roads of Glen Roy, and, of course, he knew how Agassiz had explained that. But he dismissed the inland-lake theory because there was nothing nearby "capable of damming up the waters to such heights." Agassiz's idea of a glacier dam went unspoken. So these sights too were remnants of the last continental submergence, not of an Ice Age.

Before returning to England, Lyell spent a week in Nova Scotia, where he found one final proof of the correctness of his ice-water theory. He was walking along a sandstone beach beneath a cliff wall when he suddenly noticed that a series of parallel grooves had been scratched into the beach's surface. Exposed sandstone erodes too quickly for the grooves to have been anything but recent. Lyell turned to his guide and asked if he ever saw ice in this area. Sure. The previous winter the whole basin had been frozen over with ice that had thrashed around with the tide, piling up an ice foot that reached 15 feet in places. The notion of an ice foot was news to Lyell, but he saw what it could do to sandstone, and he accepted the markings as proof that all the grooved rock in North America came from floating ice.

THE GRIMSEL HOSPICE,
SWITZERLAND, AUGUST 8, 1841...

One day after Lyell's first arrival at Niagara Falls, James Forbes arrived at the inn near the Grimsel Pass in the high Alps. Agassiz had reached the inn a few hours earlier. It had been ten years since Forbes's chance meeting with Venetz and his introduction to the glacier theory, but Venetz had not inspired him with any interest in glaciers. After dismissing the glacier theory as "a rather bold speculation," Forbes seems to have had no second thoughts. He had been back in Switzerland in 1839 and hiked across glaciers then, but in 1843 he wrote, "I cannot now recall, without some degree of shame, the almost blindfold way in which, until lately, I was in the habit of visiting the glaciers. During three previous summers I had visited the Mer de Glace and during two of them, 1832 and 1839, I had traversed many miles of its surface; yet I failed to remark a thousand peculiarities of the most obvious kind, or to speculate upon their cause."

But now Forbes had come to remark and speculate. His meeting with Agassiz the previous year at Scotland's northern tip and Agassiz's invitation had led to this reunion at the Grimsel Hospice. Agassiz was spending the summer, as he spent every summer from 1840 to 1845, camped with a team on the Unteraar glacier, studying in detail the glacier's movement and structure. The previous year a great many of the curious came to the Hôtel des Neuchâtelois, as Agassiz called his camp, and there was no reason to think Forbes would be any different.

The next morning Agassiz and Forbes went for a hike along the *glacier inférieur de l'Aar* (the lower Aare glacier). The glacier here was a true ice river, complete with tributaries. It was formed by the confluence of two other glaciers, the Lauteraar and Finsteraar. A moraine alongside the glacier marked how much farther out the glacier once had reached. Agassiz acted as guide while Forbes took the position of a humble student. Presumably the experience was similar to the times Buckland gave his geological survey of the Thames valley. Forbes was discovering how much he had overlooked during his previous "blindfold" hikes across the ice. The point was to impress the listener with the speaker's knowledge, forcing a quick surrender to the new expertise. But then Forbes pointed out something to Agassiz. He noted a "vertical stratification," alternating blue and clear bands of ice that ran across the top of the glacier. Arnold Guyot had reported such bands years before at the Porrentruy meeting, where Agassiz converted so many continental geologists, but nobody had thought the bands were important then, nor had they known what to make of the information. Everybody, including Agassiz, seems to have forgotten the report.

According to Forbes, Agassiz said that he doubted the bands penetrated into the glacier's depths. The surface of glaciers, Agassiz told him, was subject to all manner of temporary oddities. He thought he had once seen a similar banding on Mont Blanc's Glacier des Bois.

Forbes was not to be dismissed, however. They continued walking, and each time they reached a crevasse Forbes was down, peering into it and pointing out this same ribboned structure. The bands ran right up to the crevasse and, it looked to Forbes, persisted deep into the ice crack. Agassiz was not convinced. At last they found a hollow in the ice, at least 20 feet deep, and here too Forbes found his bands. Agassiz now gave up and admitted the bands were real and part of the glacier's structure.

For Agassiz, who had studied the glacier just the previous summer, the walk had become acutely embarrassing. He was the expert, and on the first day a physicist noticed something he had never seen before! It could not be. Agassiz said he thought the bands must be a new phenomenon, introduced during the previous winter. It could not have been there during his exploration in 1840.

Forbes pointed to a crevasse and asked how old it was. A couple of years, Agassiz said; anyway, certainly more than a year old. Forbes got down to study it. The bands crossed the crevasse. More telling was the way the border of the band bent just at the lip of the crevasse, suggesting a distortion brought on by the movement of the ice as the crevasse had opened up. The lines, in other words, were older than the crevasse, and the crevasse predated 1840. Agassiz had simply missed the details.

Adding to Agassiz's discomfort was Forbes's boasting. He told everybody about his discovery, adding his astonishment that such a detail had gone unobserved by Agassiz and by all other visitors. A few days later Bernard Studer, an eminent Swiss geologist and a frequent visitor, arrived at the Hôtel des Neuchâtelois. Forbes told Studer all about his discovery and again expressed amazement over Agassiz's ignorance of the facts. Adding salt to the wound, Forbes insisted that this was no casual observation. He suspected that these bands were key to explaining how glaciers moved.

Agassiz had a very thin skin. He liked the sweet praise of colleagues, but he had no taste for their sour dispute and their parading of rival egos. He did not like shared glory. He was already in furious dispute with Schimper over who had first thought of the Ice Age. Probably Agassiz could have spared himself that fight if, in his book, he had only mentioned Schimper's contribution. Agassiz had even reprinted Schimper's simple little graph of temperature changes without citing its source.

Charpentier too had broken with Agassiz. By publishing a book on glaciers a few months ahead of Charpentier, Agassiz had scooped his teacher. Charpentier never forgave him. Agassiz's defenders point out that his book gives plenty of credit to Charpentier, but Charpentier felt that he had been working on his book much longer and had a right to publish first.

These incidents could have taught Agassiz something, but it seems not to have been a lesson he cared to learn. In October 1841, after his summer on the Aare glacier, Agassiz wrote to Humboldt about his research and said, "The newest fact that I noted is the presence in the mass of the ice of vertical ribbons of blue ice." There was no mention in the letter of James Forbes, and Forbes found out about Agassiz's omission.

Forbes published an article of his own in which he stressed Agassiz's ignorance of this ribbon structure and how surprising it was to see something unknown to Agassiz. However Forbes intended these remarks, Agassiz took them as a slap of the gauntlet, and the two scientists began a long public feud, with open letters flying back and forth. Eventually Agassiz remembered, or was reminded, of Guyot's paper in 1838 and claimed priority after all. Forbes, for his part, switched from physics to glaciology and began his own annual pilgrimage to Switzerland to study its glaciers.

The issue of the blue ribbons was of absolutely no importance to the Ice Age evidence, and the entire episode could be left to historians of glaciology, but for one detail. The feud hurt Agassiz's reputation in England very badly. The days when the Geological Society regularly awarded him prizes had passed. Lyell tried to speak up for Agassiz, saying he believed both Agassiz and Forbes had independently discovered the blue bands, but England was washing its hands of Agassiz. His Ice Age theory and catastrophism did not sit well there, and it became easier to dismiss him if you could both dispute his views and frown on him for ungentlemanly conduct.

BEFORE THE FACE OF THE HUMBOLDT GLACIER, GREENLAND, APRIL 28, 1855...

Kane was wiser about the Arctic now. The previous year he had begun sending out exploration groups in March, far too early. This year he knew that neither he nor his men could do anything until later. It would have to be full summer before they tried to escape across Melville Bay, and even then the sight from the iceberg lookout that had ended their attempt to reach Beechy Island might be waiting for them again.

Once the sun was truly up, Kane made a return pilgrimage to the great glacier, his expedition's most enduring discovery. Of the Franklin party, they had seen nothing. Of the Open Polar Sea, they could say *maybe*, but even that was the same maybe reported so often in the past: *We had gone as far as we could, and in the distance we saw open water.*

But the glacier was something new. As a practical matter, Kane had found a barrier to travel, and he could report that any further overland explorations should use Ellesmere Island on the north side of the channel, and indeed Ellesmere Island did become the jumping-off point for future American polar expeditions. Robert Peary used it, and so do today's adventure travelers.

As a geographical matter, Kane had found the world's largest glacier—a finding to match discoveries by other midcentury explorers who reported Mount Everest and Lake Victoria. Besides naming the glacier, Kane had named the two points at either of its edges for two glacier experts he admired. The nearest one he called Cape Agassiz; the other was Cape Forbes.

The discovery also had something to contribute to the Ice Age debate. After almost twenty years the Ice Age theory still had not been generally accepted because most geologists still could not conceive of truly great ice. Even when their most admired peers stood before them and said they should picture Antarctica or Greenland, geologists still could not conceive of it. Whether it was due to mediocrity, or timidity, or a principled refusal to conceive of anything their own eyes had not witnessed, most scientists could not grasp the reality of great ice. For them, glaciers meant the high Alps. They did not place glaciers among the greatest, most monstrous, most overwhelming structures on the planet. Glaciers for them did not include the possibility of Melville Bay's 150-mile coast or the Humboldt's moving wall that rose almost three times higher than the chalk cliffs at Dover.

The ice that rose in front of Kane was the reality geologists had missed. If it had been in the United States, the Humboldt's snout could have covered the whole stretch of Florida's Gold Coast from Fort Lauderdale to Palm Beach and beyond. If it had been in Europe, it might have swept down the North Sea and stretched a wall along its entrance from Dover in England to Calais in France and then reached the same distance and a dozen miles more into the French countryside. It was no Agassiz dream; it was there in front of Kane, prowling across the earth, as unexpected by science as was the first dinosaur bone.

Kane had his scientific ambitions, and he knew the Humboldt glacier was news, so he studied it as closely as he could. It struck him that the place where icebergs broke off from the glacier was more ambiguous than he had first recognized. In his first book about the Arctic, Kane's description of an iceberg crashing into Baffin Bay seemed as precise as death. There, on the mainland, the ice moved with the general force of the glacier; here the ice moved with the currents of the bay. In between was a debacle that violently ripped some of the ice from the glacier and transformed it at once into an iceberg.

Standing in front of the Humboldt, Kane was not so sure he had been right. For one thing, he realized that, despite the enormous tangle of ice, leads, and bergs, the sea was not constantly in turmoil. If icebergs were calving into the sea all along a 60-mile front, the water should have been raging like Prospero's tempest. Instead, it was a mostly calm maze. A terrible maze, but not a raging one. There was no violent break between glacier and iceberg. Rather, the Humboldt propelled its ice forward "step by step and year by year," and after reaching the water, it continued forward. The glacier was not something that halted automatically at the shoreline, like an anxious beachgoer afraid to get his ankles wet. It pushed into the water and, Kane figured, kept going until the depth was great enough to float the ice. Then a new berg rose from the sea and floated off, eventually to reach a climate warm enough to melt it down.

He had been serious the previous summer when he wrote in his journal that he wanted to contribute to the geology of glaciers. Even before setting off for Smith Sound, he had sent a letter to Agassiz comparing icebergs and glaciers:

Dear Sir,

Allow me to call your attention to an extract from the book of Prof. Forbes on the Pennine Alps. "The floating masses called Icebergs in the Polar Seas are for the most part of the nature of névé, mere consolidated snow." He stated too, in the same consideration "that the occurrences of true ice is comparatively rare, and is justly dreaded by ships."

This does not, at all, accord with my own experiences of David Strait which form, as you are aware, the Chief Seat of our Atlantic Supply. Making every allowance for variances of definition, their structure was not that of névé at first, but a true ice which although verticular and sometimes opaque, had the luster, fraction and other external characters of Glacier Ice. My observations on the Coast of Greenland force me to regard the Polar Glacier as..., except in some minor and interesting particulars, based on the differences of climate, and especially the alternation of day and night, identical with its Alpine Cousins. The ice berg is merely a disruption from its mass and as such partakes of its glacial character.

Might I take the liberty of asking whether you regard the ice berg of the Polar Seas as for the most part of the nature of névé?

Plainly, the letter's tone was all wrong for a man seeking a mentor. He could have just as well ended with the question "Might I take the liberty of asking whether you are an idiot?" There was much too much telling Agassiz what he (Kane) thought and far too little overt respect for Agassiz's wisdom. Even worse, from Agassiz's perspective, all the attention was to the opinions of James Forbes, whom by then Agassiz deeply resented. As far as the record shows, Agassiz never replied.

In its blindness to Agassiz's ego, the letter was a perfect expression of the man Kane had been before his second Greenland winter. But the letter also shows how ignorant even glacier scientists still were. Readers today wonder how a question about the similarity between glaciers and icebergs could even be asked. Kane now had the evidence to set science straight, but he and his men were still trapped in a frozen world with a ship they had cannibalized beyond all seaworthiness.

If they could contrive somehow to get out of their trap, Kane would have gained a position from which he could dazzle the geologists of the world. Kane could put an end to the instinctive way the word "glacier" evoked scenes of Switzerland. The limits on the ordinary geologist's imagination could be swept aside by a rich portrait of Ice Age–strength glaciers and the ice sheets behind them. Greenland would replace Switzerland in the common imagination as the setting for an Ice Age. Or, at least, it would, if Kane could somehow contrive to escape. Otherwise, the world would just have to wait until some rescue party found the group's bones and Kane's accompanying journal.

Kane could not know it, but that spring the fate of another lost expedition was finally resolved. A search party at last located the Franklin party's remains. Franklin's crew had been dead for years, of course; dead long before Kane's first trek to find them. They had become trapped in some northern nowhere of remotest Canada and had died trying to journey overland by foot. Documents and baggage found with the bodies settled the story. Now Kane was planning a similar operation—only he was much farther north than Franklin ever sailed. There was no reason to expect anyone back home would ever hear Kane speak about the Humboldt glacier, so he prudently wrote down what he learned.

EDINBURGH, SCOTLAND, JANUARY 13, 1842...

The Scotsman appeared on Edinburgh's streets carrying a new science article by Agassiz's admirer, Charles Maclaren. At age fifty-nine, Maclaren was only two weeks away from marriage (for the first time), but apparently he had been thinking about more than romance. The latest issue carried a new idea about the Ice Age: Its huge glaciers would have shrunk the oceans.

Maclaren was one person who had truly surrendered to Agassiz's idea, surrendered in a way that our own age of the hero-rebel finds difficult to grasp. He had not surrendered publicly while holding secret treason in his heart. Nor had he surrendered passively, turning to silence and thinking about other things. He had given in like the bather on a beach who dives into chilly water and starts to swim. Maclaren became aggressive and imaginative in this new context, and he made the Ice Age theory his own. He had thought about the implications of a global ice sheet and had passed his conclusions on to his readers, who, by the way, were not educated aristocrats but were mostly Edinburgh's artisans and small shopkeepers. It was the humbly educated who sat in their homes and on the omnibuses reading Maclaren's argument: "If we suppose the region from the 35th parallel to the north pole to be invested with a coat of ice thick enough to reach the summits of the Jura, that is about 5,000 French feet, or one English mile in length, it is evident that the abstraction of such a quantity of water from the ocean would materially affect its depth."

In an age when people routinely worry that global warming might mean the melting of the polar ice and the rise of the sea level, Maclaren's association of great ice and ocean depth seems obvious; but in 1842 it was an unexpected thought. The link suggested a flaw in Goethe's speculation that there had been both an Alpine ice sheet and inundation from the sea. It also was a warning that Lyell's idea that much of North America had been underwater during a glacial (cold) epoch might not be sound.

It seems surprising that this idea first appeared in a newspaper and was the insight of an editor; however, not many naturalists were mulling over the effects of the Ice Age. Besides, *The Scotsman* was no ordinary newspaper, and Maclaren was an even more extraordinary editor. He was a hardworking optimist, a self-taught nobody who had risen to prominence through the force of his printing press. He was the kind of self-made disrespecter of privileged persons that Americans were already accustomed to but who were still something of a novelty in England and Scotland. Indeed, Maclaren was enthusiastically pro-American and pro-republic. He wrote the general entry about the United States for the *Encyclopaedia Britannica,* which, it was said, was the only entry about America that did not have to be rewritten for the encyclopedia's U.S. edition. Britain, of course, had its share of self-made achievers, but Maclaren was unusual in the way he blushed neither before peers of the realm nor before the most educated moderns. He was confident that he could face even the best of them as equals if he worked at teaching himself. So he taught himself.

He attended Jameson's lectures on natural history at the University of Edinburgh, and he frequently explored the Scottish countryside to study its geology. As early as 1828, two years before Lyell's first book, Maclaren had published an article in *The Scotsman* about alluvial phenomena near Edinburgh. Other geology reports followed, based not on interviews with geologists, but on his own explorations. In 1839, he published a book on the geology of Scotland's Lowlands around Edinburgh. He wrote an occasional "Notes on Science" column for *The Scotsman* that covered physics and astronomy as well as geology.

He had a good mechanical mind, probably better than Agassiz's. The relation between the size of the ocean and an ice sheet is simple and mechanical. Agassiz imagined complex mechanical events like the rising Alps, cracking glaciers, and sliding boulders. Maclaren's thoughts were more basic: Water cannot be both frozen and liquid. More of one implies less of the other.

This kind of simple mechanics had led to Maclaren's first big science success. In December 1824, *The Scotsman* had begun publishing a series of articles about railways in which Maclaren described how high-speed (20 miles per hour) rail travel was possible. Maclaren spelled out the physical principle that would make railroads the dominant source of freight transport: Low friction between rail and wheel allows relatively weak horsepower engines to pull great loads. These articles appeared very early in the course of things, for the first modern railroad is generally dated at September 27, 1825, and tradition says the railroad era began on September 15, 1830, with the opening of a railroad connection between Liverpool and Manchester. Maclaren's articles heralding the new transportation method was one of the first roosters to crow at the industrial dawn. The series was reprinted in America, and in French and German newspapers as well. But although Maclaren's reports caught the popular imagination, expert opinion was more skeptical about the low-friction concept. Maclaren had based his analysis on some eighteenth-century experiments that, by 1824, most scientists had long since forgotten. As we all know, Maclaren turned out to be the one on the money.

It is not that the experts were fools, any more than the geologists of the 1840s were. It is just that most people, experts included, like to hedge their bets. They don't surrender to the new, and therefore they do not make the new their own. Maclaren did not hedge. Politically, he was more Whig than the Whigs. Personally, he was more American than many Americans. Intellectually, he had begun thinking more creatively about the Ice Age than even Agassiz. This total approach made him suspect, of course, among all those who had placed their bets on older ideas. There was a story that, in 1817, soon after he and a friend had founded *The Scotsman*, a laird had been seen visiting the humble cottages of his district, telling the villagers he was most troubled to hear that some of them had been seen reading "that incendiary paper, *The Scotsman*" and that if they would cease, he would supply them with other newspapers at his own expense.

Maclaren's hard work and intelligence eventually gained him a solid reputation in Edinburgh. He became friendly with Jameson and was, in 1837, chosen as a member of the Royal Society of Edinburgh, but his background was always available as a weapon in the hands of opponents. Scientists in the 1840s liked to think they were more open-minded than the pre–industrial age peers who had first opposed Maclaren, but they did not respond much more thoughtfully to the notion of the shrinking oceans. Maybe they reacted even more poorly. *The Scotsman* had prospered despite the opposition of the lairds, and the success of the railroads quickly silenced skeptics, but it would be a long time before the link between great water and great ice was more than a rumor enjoyed by the more thoughtful shopkeepers of Edinburgh.

A month after Maclaren's articles appeared, Roderick Murchison gave his presidential address to the Geological Society of London, and he refused even to accept the Swiss glacier. He told the assembled geologists:

> *Once grant to Agassiz that the deepest valleys of Switzerland, such as the enormous Lake of Geneva, were formerly filled with snow and ice, and I see no stopping place. From that hypothesis you may proceed to fill the Baltic and the northern seas, cover southern England and half of Germany and Russia with similar ice sheets, on the surfaces of which all the northern boulders might have been shot off.... So long as the greater number of the practical geologists of Europe are opposed to the wide extension of a terrestrial glacial theory, there can be little risk that such a doctrine should take too deep a hold of the mind.... The existence of glaciers in Scotland and England (I mean in the Alpine sense) is not, at all events, established to the satisfaction of what I believe to be by far the greater number of British geologists.*

GENEVA, SWITZERLAND, AUGUST 11, 1845...
Continental geologists were different. Mostly they did think a great glacier had covered the Swiss valleys. In England the pressure against the glacier idea was so great that even Buckland had grown silent, but in Europe the pressure worked in the contrary direction. There it was the skeptic like von Buch who was made uncomfortable. He had come to Geneva for the Swiss Society of Natural Science's annual three-day meeting. Geneva in August seems the perfect place to be. The climate is mild, the mountain backdrop magnificent, and the lake enchanting; but von Buch was so distressed by this meeting that he fled it early, heading back north to Zürich.

Murchison, however, had been wrong in his insistence that the Swiss glacier was the thin end of the wedge, inevitably bringing the whole Ice Age behind it. By 1845, most geologists in France, Switzerland, and Germany took it for a certainty that a monster glacier had buried the whole of Switzerland, stretching to the Jura in the west, Italy in the southeast, and Austria in the north. Yet they did not believe in an Ice Age.

The importance of the Scottish disaster was now clear. With the exception of Maclaren and a few other amateurs, the rejection of Agassiz's and Buckland's evidence for a Scottish glacier had put a halt to work promoting the Ice Age theory. Something new was needed if it was to be revitalized. In the meantime, the Ice Age idea had developed a limp. Humboldt had written to Agassiz in the summer of 1842, "'Grace from on high,' says Madame de Sévigné, 'comes slowly.' I especially desire it for the glacial period and for that fatal cap of ice which frightens me, child of the equator that I am."

Fatal cap of ice which frightens me. There has not been much in human history that the fear of death could not slow down. Yet the idea had not disappeared entirely. Perhaps Murchison had been right that, over the long term, acceptance of the Swiss glacier would eventually mean acceptance of the Ice Age. If he was correct, these continental geologists, especially the Swiss ones, were playing the role enacted by Irish monks during the Dark Ages. They kept knowledge alive after its energy had elsewhere been expended. For an antiglacier man like Leopold von Buch, the meeting of the Swiss Society was startling evidence of just how alive and ruddy the great-glacier idea had grown. Agassiz was there, reporting on his investigation of the Aare glacier during the previous three summers. Charpentier was there too, along with Ignace Venetz's son, and the two of them conducted a discussion about the importance of the early observations made in the Rhône valley.

Von Buch tromped out. In Zürich he looked up a long-time member of Agassiz's team, Arnold Escher von der Linth, and made a strange request. He wanted Escher to lead him through the Alps of central Switzerland; however, Escher was not to say anything at all about glaciers or glacier actions. Escher agreed, and together they made their way to the Rhône valley and the Valais.

Von Buch was seventy-one years old, but he was a hiker and climber of many years' training. A diarist in 1837 had described him in the Jura:

> *I shall only say that the old von Buch made us run like skinny cats. This man who appeared to be able to see only a foot's distance in front of him, who walked with bent knees,*

> *wore large pumps with loose buckles, set his foot in a trembling manner and proceeded always straight in front of him with his eyes on the ground, made us walk so fast that we, young people, could barely keep pace with him....He stopped from time to time, however, to break rocks and examine them. He put in his pocket those which appeared interesting to him, after having carefully wrapped them in paper.*

Eight years later, von Buch might have slowed a little, but not by much. He went up and down the slopes with Escher, where there was plenty to see. In central Switzerland, moraine evidence is particularly strong, and moraines are the most obvious product of glaciers. The road up to the Simplon Pass goes by an ancient moraine so large that the glacier that made it must have stood at least 200 feet high. On the flanks of the Rhône valley, high up on the slopes, far from any existing glaciers, stand a series of moraines. Von Buch might always protest the transportation of erratics by glaciers, but how to explain these moraines?

And then von Buch went on his way. Escher had kept his word and had been silent about glaciers. Von Buch said nothing either, although his vocal condemnations of the Swiss-glacier theory did end. His surrender appears to have been solely of the useless, silence-producing kind that is liable to produce sneering jokes in private but no useful insights for the public.

Years later, in the early 1850s, when von Buch was nearing eighty, he was still hiking in the Alps and still seemed a doubter. This time he traveled with Roderick Murchison, whose antiglacier sentiments were even more vocal than von Buch's. The pair of them came across a house-sized, erratic boulder, and, like Moses looking for water, von Buch struck it with his cane. Where, he demanded, is there a glacier that could have carried this block and left it here?

Murchison, who was as ignorant as an equatorial babe of Greenland's great ice, smiled at this telling question.

Part V

Thrust Home

RENSSELAER HARBOR, GREENLAND, MAY 17, 1855...
The infernal spring had softened to mere purgatory. Although the wind could still cut, the temperatures were no longer merciless. The sun had climbed so high that it had stopped setting. Yet the ice was still sound as a marble floor and still held the *Advance* as fixedly as a crocodile holds a baby. Of course, even if the ice gave way, the ship was useless as a sailing vessel. The party had burned too much of it for warmth through the monstrous winter and the early, killer spring. Even if all the ice turned into a memory, the ship was no longer sturdy enough to endure a voyage and was no longer the way home.

Near the half-eaten ship, the entire party—including those too feeble to do any work—had assembled with their three longboats and their sleds. Each of the boats was decorated with a linen rag that had been colored with red and white stripes and white stars on a field of blue. With these reminders of the American flag to guide them, the strongest men in the party began dragging the boats toward the west. It was a terrible weight, of course, even for healthy men. The boats were raised partially on sleds to ease their portage, but the contrivances were still clumsy and demanded the concentrated resolution of the physically ruined men. They were abandoning ship.

For weeks Kane had prepared the men for this date, announcing well in advance that it would be the day certain for their final attempt at retreat. Those who did not die in the trying would get home at last. To survive, they would have to cross the 1,300 miles to Upernavik by foot and longboat. Once the men had dragged the boats to open water, they would start rowing. Kane himself was in the best shape—perhaps because of his rat protein—but many in the party were too weak to carry even themselves to the ice floes. While the fittest men walked, Kane used the dogsled to haul those who were too sick, as well as the party's luggage, to a depot near the ice edge. He could cover in a few hours an icescape the men would need weeks to cross.

Adding to the venture's risks was the general unseaworthiness of the longboats. Christian Ohlsen, the carpenter, had done what he could, but keeping them afloat would demand constant refitting. Yet the men made a good show of heartiness. The surliness and contempt they had once shown Kane's leadership was gone now, and they moved with a disciplined will. They would have to drag the boats almost 100 miles across the ice. The first day they managed 2.

That first night Kane led the men back to the brig. In the morning they returned to the boats, dragged them a few more miles, and then hiked back to the brig for one more night of hot food and good shelter.

It was only on the morning of May 20 that most of the party said a final farewell to the ship that had seen them through two sunless winters. Left behind on board were all the natural history specimens they had collected; the brush, bones, and bugs that had once seemed so important now looked like useless weight. They did carry the ship's figurehead, a log-sized carving nicknamed "the Fair Augusta," that, in a last extremity, could be used as firewood. Kane gave the men a pep talk, saying that they faced great challenges but could succeed if they obeyed orders and brought energy to their task. He knew that none of the men had any physical energy left.

Kane continued to use his wretched dogs to haul people and food to the depot across the ice. In Eskimo style, he wore slit goggles to protect his eyes as he sledded across the star-bright ice. The frozen route was difficult and monotonous besides, so Kane found it both exhausting and boring to travel mile after mile across the ice flats. The icebergs trapped out in the sound had become too routine for them to inspire any curiosity now. Land to the south sloped upward, steeply in places, but it too was a frozen wilderness, with black stones locked into its white ice. At one point, when he could stand the tedium no longer, Kane tried to follow the coastal ice belt, but a large chasm in the snow sent him back to the safety of the open ice. Although they were awful, the ice flats were safest as long as they stayed solid.

Occasionally, Kane returned to the brig to fetch more supplies, especially food. The abandoned galley returned to life as he melted some of the frozen pork and baked bread. Judging the store of dried apples "still eatable," he gathered them as up well.

Day after day, out on the ice flats, the men continued dragging the boats across the tedious ice. Healthy men develop a rhythm and momentum when they do this kind of work, so that once they are under way they can practically jog along with their heavy load flying beside them. But these men were not sturdy enough in mind or body to do anything so hearty. The dogs too were worn out from hardship and labor. Once in a while they would just give out and force Kane to camp wherever he happened to be.

Twelve days after beginning the retreat home, Kane packed up the last of the brig's supplies—bags of "flour pudding" and pork fat. Then he too put the ship to his back and went to face however long the retreat would take. Returning to the depot, he received a new shock. He had been crossing the frozen desert when he reached a kind of promontory that allowed a miles-deep view of the flat ahead. Previously the view had shown the monotonous wilderness, but now the scene had turned cruelly interesting. The ice was beginning to melt. The snow had changed from Arctic white to soupy gray—taking on the leaden look of a snowball as it melts in your glove. Pockets of water seeped through and turned the flats into a kind of bog. The breakup of the ice had been last year's unanswered prayer, but it would be this year's nightmare. Before, the *Advance* could have forced its way through the floes and made good an escape. Now, with only flimsy boats, the open floes would make portage impossible, yet they would also block any hope of rowing through the ice pack. Kane suddenly understood that the retreat depended on winter, but even in northern Greenland summer was coming.

There was nothing to do but keep trying and hope they escaped before too much of the hated ice disappeared. Kane made it to the depot and gave orders for the men, no matter how sick they were, to prepare for instant flight. Then he left to beg help from the Eskimo village at Etah.

The surviving dogs had done valiant work, but they were too feeble and too few for rapid escape. With the crumbling of the ice pack, Kane needed more and better dogs to speed the portage. The Eskimos had the necessary dogs, but a request of the sort Kane had in mind would put a great strain on the people's fate. They had no dogs to spare.

Etah was about 75 miles south of the Greenland coast and sat beneath a black cliff. Emerging between two pillars in the cliff, towering over the people and their homes, was a glacier that, like the ones Agassiz had studied, gave off a stream of water at its snout. The creek emerged from an archway at the glacier's base and provided the village with a lake. Etah's lake even had fish in it, although Kane reported that, surprisingly, these Eskimos did not know how to fish and never ate them. Even though time was essential, Kane was ever the glaciologist, and he studied the meaning of this river in the ice. He reported that, even in darkest winter, when all surface water is hard as stone, the running water of the Etah glacier could be heard beneath the ice. "This fact is of importance," Kane wrote, "It shows... these great Polar glaciers retain a high interior temperature not far from 32°, which enables them to resume their great functions of movement and discharge readily when the cold of winter is at an end, and not improbably to temper to some extent the natural rigor of the climate." Kane could not know it, but these under-ice rivers flowing through the glaciers were the source of the strange long mounds (called "eskers") that Lyell had seen in Sweden and again near Toronto.

When Kane sledded into the village beneath the glacier, he had no bargaining power beyond his own human need for help. The Eskimos listened and then let Kane trade ruined dogs for better and take a supply of walrus meat as well. Then, with his new team in harness, he sledded back toward the men who were still dragging their boats across the weakening ice.

It seems a marvel that Kane could always find his way back to the men doing the portage work. They were small dots on a large canvas, but he seems to have found them repeatedly and without special difficulty. This time he found them struggling through the wet snowdrifts. He reached them just in time to endure the worst gale they had yet seen in Greenland. The summer solstice was only a few weeks away, yet here was a wintry storm with winds ferocious enough literally to blow the new dogs right out from their harnesses. Seeing that, the men threw themselves down to hug the ice. As they lay there the terrible wind ripped their fur jumpers from their backs, exposing them to the howling Arctic elements.

The instant a lull came, they were up and calling the dogs. The animals reappeared, and the group struggled desperately to haul the boats to the modest shelter of a nearby island. The storm returned and buried both dogs and men under a snow hill. Again Kane risked the fate of Cuvier's mammoth, but the men arranged themselves "spoon" fashion to share their warmth. They lay together, front to back, with their legs tucked up so that they lay snug as a stack of spoons in a chest drawer. From above, the party was just another pile of drifted snow, a hill growing higher in the storm. From inside the drift, however, they were living bodies giving off heat and moisture, like frogs hibernating through the worst of times.

Whatever relief they felt when the storm finally passed could not endure, for the ice had become still more dangerous. The ice marshes were giving way to open pools. Now as they moved they had to pick their way between places where the ice was too weak to support the portage, or even places where the ice was altogether gone. Inevitably, there came a moment when they picked badly and the ice gave way. A sled supporting one of the boats went into the drink, taking six men in behind. The boat too was almost lost, but the men who were still standing managed to save it and rescue the splashed-through members as well. The party was now soaked, without adequate furs, and exposed on the weakening ice.

Fortunately, when life is so desperate, it is also quite simple. There was only one thing to do, and they pushed on, picking and pulling their way through the ice marsh so that twenty-four hours after the storm ended, they finally reached the edge of the floe. They had done what they had to do—dragging themselves and their equipment 100 miles across the late-season ice. It had taken them almost a month.

Making prospects even brighter, the Eskimos from Etah suddenly joined them, carrying meat and bringing rested dogs. The party's official story to the local people had been that it was going on an extended hunt, but the villagers may have seen through that. Even white men do not go on a hunt with people who are too weak to move. Everyone from the village had turned out for a farewell.

Still, the trek to the ice edge was not quite complete. The sick members and all the party's food stores were still back at the depot. Taking advantage of the fresh dogs, Kane set off on one final ride back to the abandoned ship to fetch whatever last food he could find. On this last, long sled ride, Kane saw how the late season was changing even the coastal geology. The black rocks that had seemed locked forever into the ice were producing remarkable effects that even Agassiz had not imagined. Freezing water in the icy cracks shot the rocks upward, catapulting them with a loud bang toward the heavens, so that as Kane and his dogs slid by they watched and heard stones leaping like popcorn kernels on a hot stove. From time to time, one flying stone would land on top of another, forming a pile of teetering rocks of the sort that make even the least geologically alert of passersby wonder how in the world nature ever laid things that way. They were a miniature version of the blocks of Monthey that Agassiz had seen near Bex.

Kane's pleasure could be only an intermission. It ended completely when he returned to the ice floes and learned that the ship's carpenter, Ohlsen, had collapsed from exhaustion. Ohlsen had noticed the ice giving way under one of the boats, and he sprang forward to save it, shoving a metal bar under the boat to keep it afloat while the others worked to drag it back out of the sea. Ohlsen put more energy into the struggle than he had to spare. When Kane found him, he was lying down, completely spent, with no strength left. Three days later, while Kane was scouting for a passage home through the ice floes, he learned of Ohlsen's unexpected death. To die of exertion seems incredible and shows how close to the border between life and death all these men had strayed. Ohlsen had been one of the stronger ones, yet really he had become only a rusted shell, with no capacity to recover from sudden work. For these men, action only hastened ruin.

ABOARD THE SOUTHAMPTON RAIL LINE, ENGLAND, SEPTEMBER 1846...

Lyell traveled easily across the British countryside on the new rails that stretched between London and the English Channel. England's most distinguished scientists were taking this speedy new route, opened in 1840, to that year's meeting of the British Association for the Advancement of Science. The glacier matter might be coming to a head again. Lyell had nothing new to add, but two other figures expected to attend were both about to publish important work. Edward Forbes (no kin of Agassiz's rival, James Forbes) was readying a paper that described fossil evidence that England had once been much colder. It was a great challenge to the imagination, the idea that this countryside that Lyell whizzed across so comfortably by rail had been as cold as the upper Canadian regions where Sir John Franklin had recently disappeared.

Agassiz was coming to the meeting as well, stopping off before traveling on to the United States. He had spent the past few months in Paris, where he had finished the first volume of his new book, *Système glaciaire,* which summarized all that Agassiz had learned about the workings of glaciers. Agassiz was sure to take the Forbes paper as independent corroboration of his Ice Age theory, despite the fact that Forbes proved only that the period had been "glacial" (i.e., cold), not that there had been a glacier. But Agassiz was such an enthusiast, he could be relied upon not to let that difference slow him down.

Sure enough, when Agassiz arrived in Southampton he wore the marks of public triumph. Privately, he had met disaster. Cécile had left him, returning to her family's home in Germany. She took their daughters with her, leaving one son in Louis's care. Divorce was out of the question, of course, yet reconciliation seemed a psychological impossibility. Neither one had been willing to press very deeply into the other's terrain, and finally they shared no common ground.

The European public, however, still cheered Agassiz and thought him marvelous. The Agassiz papers at Harvard University include a wonderful proof of his renown: a letter mailed by an admirer in 1845 containing a beautiful, hand-drawn map of the erratic rocks in the Rhône valley. The sender, one Edouard Collomb, seems not to have known how to find Agassiz, for the letter was posted simply to "Monsieur L. Agassiz, Member of the French National Institute, Neuchâtel (Switzerland)." Even without an address, the Neuchâtel post office knew where to deliver it. Still more impressive had been the torchlight parade and student serenade given in his honor before he left Neuchâtel. He was expected to be gone for about a year—traveling to Paris, Britain, and then the United States. Of course, people regularly left Neuchâtel for a prolonged absence, but in Agassiz's case the townspeople had organized this ceremony to say they loved him and he should not forget to return.

In Paris Agassiz's work had gone well. (His book was to be published in Paris, not Neuchâtel, because Agassiz's private publishing house had finally run out of funds.) Lyell considered the metaphysical interpretation that overlay all of Agassiz's science absurdly old-fashioned, yet the book was an extremely modern scientific achievement. It had two coauthors—Arnold Guyot and Edouard Desor—and presented the findings of a whole team of researchers. Very few science books of that era were team efforts. The book was only the first volume of a planned three-volume work. The project's ample ambition was shown in its subtitle: *Research into Glaciers, Their Mechanisms, Their Former Extent, and the Role They Have Played in the History of the Earth.* This first volume was confined to describing glaciers in general: their appearance, the causes of their movement, and the climatic and geographic conditions that create them. Compared to this effort, Lyell's most recent book, describing his travels in North America, was hardly serious science at all.

Inevitably, Lyell, Forbes, and Agassiz joined up at the conference. Agassiz knew Forbes from previous meetings and seemed to like him. They had spent much time together during the Glasgow meeting, and after Agassiz returned to Switzerland from his Scottish tour, Forbes sent him a flattering letter saying, "You have made all the geologists glacier-mad here and they are turning Great Britain into an ice-house." But now Forbes's ideas were much closer to Lyell's than to Agassiz's.

Especially distressing to Agassiz was Forbes's fossil evidence that the same species of shellfish had lived in the waters off Britain's coast before and after "the glacial epoch" (whatever that might have been). For Agassiz, the Ice Age had been a dead season, erasing one creation and preparing the stage for another. Ever since his Neuchâtel Discourse, Agassiz had had to face skeptics who disbelieved in the Ice Age, so the biological effects of the event had not been much disputed; but in Forbes Agassiz faced a critic who believed, at least, in the great cold, but not in the destruction of all life during that time.

Although only thirty-one years old, Forbes had already become natural science's foremost student of undersea fossils. He had done extensive dredging in the Aegean and in the waters around Britain. Oyster fishermen had, for centuries, developed techniques for dragging buckets along the bottom of the sea to fetch up a seafood harvest. More recently, machinery had been devised to break apart and haul up sandbars to make harbor navigation easier. Forbes was the first man to turn these techniques to science. He hauled up shellfish from the sea bottom to catalog their species, and he could break through the seafloor to search for fossils.

Dredging had established the cold time. Forbes had discovered that Arctic species had once lived right off the coast of Britain, implying very strongly that Britain had been as cold then as the Arctic is now. Forbes also reported that many of the same shell species that had lived off Britain before the cold had returned and still lived there in 1846. Instead of becoming extinct during the cold, they had gone south and then migrated back, just as Arctic species had moved up and down the latitudes.

This finding bore directly on the long-standing rivalry between Agassiz and Lyell. Agassiz still insisted that a catastrophic Ice Age periodically wiped out all life on earth, while Lyell still denied catastrophes any role in natural history. Lyell taught that species were created and became extinct at a steady rate. Practically, this dispute meant that if a list of preglacial species were drawn up, Agassiz believed there would be no modern species on the list, while Lyell believed that a certain percentage of modern species would be there. The farther back in time the list went, the smaller the percentage of modern species. Although neither Agassiz nor Lyell had had decisive evidence for their positions, Forbes's findings seemed to make Agassiz's absolutism untenable.

Agassiz's first reaction was to deny that the preglacial fossils had been the same as the postglacial ones. To make his case, he pointed to differences between modern and ancient shells. Forbes and Lyell agreed that the differences existed, but they claimed the shells still belonged to the same species. They were just different varieties of one species.

To nonspecialists, the argument over variety versus species can sound precariously metaphysical, but to these men, it was a crucial debate. If the shells were varieties of one species, then they were genetically linked, and the connection between them could not have been wiped out in an Ice Age. If they were separate species, then they could not have any direct link. (Neither Agassiz nor Lyell believed in evolution and did not believe there were any genetic links between separate species.)

This kind of dispute might have been left to the lounges of Southampton, but neither Agassiz nor Lyell nor Forbes was an armchair intellectual. They boarded a boat and set out into the Channel to have a look, hauling up bucketsful of mollusks to see what they could see. Mollusks are shell animals like snails, oysters, clams, and conches. When dead, their soft bodies decay rapidly, but their shells can last forever. Agassiz was basing his dispute on subtle variations in shells. Oyster shells, for example, have ridges that can be closer or farther apart. The shells themselves come in various sizes and shapes. Some are flat, some seem almost furry. What do these differences tell us about their basic natures? Are they separate species, the way leopards and lions are distinct, or are they mere varietal differences, the way spotted leopards and black panthers are varieties of the same species?

Rocking out in the Channel, hauling up their buckets of shells, the three men argued these points. Agassiz was adamant at first, and since the distinction between species and variety depends as much on interpretation as on observation, the debate seemed perhaps impossible to settle. But, as the sailors stood by, these three gentlemen continued getting themselves wet and their hands dirty. They pored over the matter dredged up from the sea bottom, and then there it was.

One of the varieties hauled up onto the deck matched the preglacial variety that Agassiz had said was another species. Agassiz had reached one of those terrible moments when a scientist must choose between his eyes and his beliefs. Nonscientists think it ought to be easy to choose and that only pride can get in the way, but it is really easy only for technicians. If anything is at stake beyond the fact itself, the challenge can be almost unendurable. There are heroes who can do it, just as there are heroes who can drag themselves across north Greenland. Buckland had been such a one, abandoning his belief in the geological evidence of Noah's flood and then giving up on the diluvium altogether to accept the Ice Age. But most scientists are not heroes. Lyell had looked for a moment like he might be one, but then he abandoned his recognition of moraines and returned to the drift and his anticatastrophe principles. Now it was up to

Agassiz to show his stuff. With shells all around, he could not hold his ground. He wobbled, offering a bitter joke that he would drop half his objections to the doctrine of surviving species if Lyell would drop half his doctrines on percentages. Lyell was not amused and years later was still grumbling about the incident. It was not yet clear whether Agassiz had assigned himself a new task—finding a compromise theory of creation and extinction—or whether he had merely gotten beyond an awkward moment.

HALIFAX, NOVA SCOTIA, EARLY OCTOBER 1846...
Agassiz seemed unchanged. As enthusiastic as ever, when his ship first anchored in the New World, Agassiz immediately disembarked and hiked toward the nearest hill. From its top he had a fine view of Halifax Bay, one of the largest and most protected harbors on the Atlantic coast. Although it was a splendid scene, Agassiz was still more pleased by what he could see at his feet, the telltale "*line-engraving* of the glacier." The rocks of Nova Scotia were polished and grooved with parallel furrows and scratches. Then it was back to the harbor, where his ship was taking on supplies before sailing on to Boston.

The city that greeted Agassiz was not the one we know today. This was Boston before the Irish immigration and before it had become a city of universities. In 1846, Boston was the financial center of America's industrial revolution. The textile mills polluting the Massachusetts countryside were owned by Boston investors, Puritans with dreams who used their money to improve Boston society without adding to its gaiety. Agassiz was one of a series of European savants brought to America by Boston money to lecture and see the new democracy. Lyell's visit a few years earlier had been under the same sponsorship.

The New England foliage was in full splendor as Agassiz walked to Pemberton Square, where one of the founders of the Massachusetts cotton mills, John Amory Lowell, welcomed the professor into his house. Lowell was no Babbitty businessman. He was truly interested in natural history and had become a patron of the sciences and arts. He was used to meeting scientific men, whereas, on his side, Agassiz was accustomed to commercial men with larger interests. He had dealt with exactly such figures every day in Neuchâtel.

Agassiz planned to make a whirlwind tour of the northern United States and then return to Boston to present a series of lectures on the "Plan of Creation in the Animal Kingdom." His theme, one that was immensely popular with his American audiences, was on the spiritual underpinning of material reality. A lecture series like this sounds, to our ears, unscientific, possibly even antiscientific; however, materialism—the doctrine that matter and its properties can account for all of existence—had not yet conquered science. The antiscientific element in Agassiz's talks was that he knew there was something wrong with his premise of repeated, separate creations and extinctions, but still he preached it.

The lectures, of course, were a success. Boston loved Agassiz's accent, his irreverence, and his knowledge. Americans, especially Bostonians, so loved Agassiz that they even cheered the lecture he gave—*in French!*—on glaciers. Boston's enthusiasm for the Swiss professor was fully requited. Agassiz fell in love with America in just the way that Americans believe visitors ought to love it. He loved its energy, its optimism, and its contempt for the past. "Their look is wholly turned toward the future," he wrote to one friend. The land was perfectly in tune with his own steam-engine energy, confidence, and ambitions for tomorrow. But there was a further side to Agassiz's love that Americans would not have appreciated so much. He knew he was bigger and better than anything the United States had to offer. There had been no man of his scientific knowledge and breadth in America since the death of Benjamin Franklin. Agassiz was not the classic immigrant who had fled an Old Country that he despised. He had loved Neuchâtel too and had turned its so-limited promise into more than anyone could have expected. The United States in 1846 was a giant-sized Neuchâtel. It was an elephantine variety of the Swiss town but still part of the same species: a society of second-class learning and first-class ambitions. Both were places where Agassiz could immediately seize and hold without contest the title of leading local scientist.

During his first month in the United States, Agassiz met many capable American scientists: James Hall, the geologist who had escorted Lyell about Niagara Falls; Asa Gray, a botanist at Harvard; and James Dwight Dana, the country's rising geologic star. The leading American scientist of that day was Benjamin Silliman, a chemist, geologist, and editor whose classes at Yale had been a fountain of American scientists throughout the first half of the nineteenth century. When Agassiz arrived in New Haven for a visit, Silliman was so delighted that he turned his class over to the guest, who gave an impromptu lecture on glaciers.

Most travelers to America did not see the country's deferential side, but for Agassiz, it was one of America's great charms. He enjoyed the company of able men who immediately ceded him alpha status. There was none of the head butting he had endured while dredging off Southampton. In Boston he soon became friends with the local giants—Ralph Waldo Emerson, Henry Wadsworth Longfellow, Oliver Wendell Holmes—none of whom dreamt of challenging him in his chosen area. In the midst of this applause, he was expected back in Neuchâtel. Instead, in February 1847, his assistant, Edouard Desor, arrived from Switzerland, and soon the rest of his team followed. Agassiz had come to stay. He had many projects in mind, and he would think of more, but he was not reworking the problem of species that had survived the Ice Age.

CAPE ALEXANDER, GREENLAND, JUNE 18, 1855...
The retreat party distributed gifts for the whole of Etah village, trinkets mostly, but Kane did surrender his surgical amputation knife and turned over all but two of the surviving dogs to the villagers. The party kept the lead dogs, however, partly for sentiment and partly as an emergency food supply. "Meat on the hoof," McGary called them.

The party tried to set off at midnight, but the sea was too rough, and the men soon returned to shore. Late the next afternoon they did put to sea and left the Eskimos of Etah behind them. At first the water was as smooth as a garden pond, but quickly it grew choppy and showed them the dangers they were facing. One boat, the *Red Eric,* was swamped and its crew of four tossed into the ice soup. The men were rescued and their boat righted, but its precious cargo of food was lost. Meanwhile, a second boat was taking on water faster than its passengers could bail. Adding to the distress was the weather. It began to rain, and by the morning of June 22, that drizzle had become a snowstorm.

The snow fell steadily as they rowed to a small landing point called Northumberland Island. The place was a miniature Greenland. Its cliffs ringed the coast, while the central plateau was covered by a miniature ice sheet that fed seven different glaciers. Especially remarkable was the way these glaciers hung over the cliff side. Large strands of ice spread from a frozen center like toothpaste squeezed from a tube. The hanging glaciers made a constant noise as they rattled against the cliff sides. From time to time, one would give off a loud boom as part of an icy strand snapped and fell to the base of the escarpment. The Northumberland coast was littered with boulders and debris carried down by these glacier cascades.

The greatest of the Northumberland glaciers reminded Kane of Agassiz's Aare glacier because it had two tiers. The first tier came almost straight down, sliding for some 400 feet along a slope of perhaps 70 degrees. Half a mile of flatland carried the glacier toward its second tier, where the ice dropped another 400 feet, this time along an escarpment with a slope of about 50 degrees. The glacier was half a mile wide at its base and 200 feet thick. Astonishingly, along its whole length, from the point where it spilled over the island's cliff tops to the shore moraine, there seemed to be no breaks or crevasses marring its form. It was as though the glacier were as pure and viscous as the goo gurgling from a giant bottle of molasses.

Just as remarkable as the Northumberland glaciers was the way Kane, desperate and near starving, still took note of the island's ice wonders. With the responsibilities of the retreat all around him and the prospect of Melville Bay's icebergs ahead, Kane still took time to sketch the hanging glacier and to illustrate the lines of viscous movement that he could see even at the glacier's top.

The Northumberland hanging glaciers flaunted the fundamental difference between Ice Age glaciers and Switzerland's valley glaciers. The Northumberland glaciers, for all their size and spectacle, were froth bubbling from an interior ice cauldron. Switzerland was all froth and no beer. No wonder that, before Kane, almost nobody could imagine that ice could spread everywhere, whereas, after him, nobody could remember what the problem had been. People before Kane had tried to picture a valley glacier getting longer, maybe wider, and stretching farther. It was as difficult to imagine how these glaciers could cover the world as it would be for an inland people to imagine a "Water Age" in which rivers had managed to cover entire continents. But after Kane, everyone understood: Forget about rivers and picture an ocean. The heart of the Ice Age is not the glacier, but the central ice sheet. Ice Age glaciers had been just little bays poking beyond the ice ocean's shoreline.

Beyond Northumberland, the party crossed Murchison Sound, named for the same Roderick Murchison who had been scoffing at the Ice Age theory for almost twenty years now. It had been on this sound's frozen water that Kane and McGary had abandoned their effort to go for help at Beechy Island. The water was open enough now to make progress possible, but already the group faced supply problems, and Kane decided to cut the daily allowance even further. Each man was allotted six ounces of bread dust and a walnut-sized lump of fat. The two ingredients were cooked into a broth and consumed in sips throughout the day. The party also boiled kettles of iceberg water into a hot tea that provided the illusion of nourishment. Rations like these were not enough to keep the party functioning, and the men began to run down like watches with broken mainsprings. Dragging the boats across the ice became impossible. They hadn't the strength. They were no longer strong enough to do the work they had been doing only a week earlier, when they had hauled their equipment to the open water of Cape Alexander. And being so weak, they could not chase food. Kane had heard from the Eskimos that there were plenty of birds to be had nearby, but finding the birds meant dragging at least one boat across a portage, and the party could not do the work. So, because they were starving, they had to starve some more.

It was time to taste dog meat. The surviving sled dogs had carried the party through many hard places, but it was time to be as unsentimental as the Eskimos. When starvation is imminent, Eskimos kill their dogs. The party's hour for that kind of desperate toughness had arrived. Only they proved not tough enough. Kane thought of the deed, but he put off the order. Bread dust today, dog feast tomorrow, and worry about Melville Bay when it comes.

The sudden appearance of eider, thousands of them, then spared the dogs for real. The party had reached the birds' breeding grounds, and all the men joined in a great egg hunt. Eider protect themselves by breeding in the impossibly remote and frigid north, so when hunters appear who are undeterred by either the distance or the cold, the birds are defenseless. The party stayed for three days, and in that time they gathered almost 4,000 eggs. A snowstorm raged about them during the whole hunt, but they ran and stuffed themselves like children put in charge of a candy store. For the dogs too it was a grand time, and they ate without worrying over how close they had come to the wrong end of a fork.

When the storm ended the calendar said it was July 4, and the men celebrated Independence Day before setting off. They poured eider eggnog and even added a dash of alcohol that had somehow survived two years in the north. Then they rowed on to an island where, it seemed, the water turned into ice. They had outpaced the season and now could do nothing but wait and hope for melting. It was impossible to know whether the ice would ever thaw, although a nearby glacier offered a good omen. The glacier's snout was 7 miles wide, with a base that rose hundreds of feet above the bay. Water emerged from its front, and from one point in the top of the glacial face, Kane could see a giant waterfall as meltwater splashed down. If water could emerge from this ice, why not from the frozen sea around them?

Luckily, they were trapped in a place where the feasting was good. Uncountable numbers of lummes nested here, and these birds were as unskilled as eider at protecting themselves. The men fed daily on eggs and meat, restoring themselves to working condition. For the party's health, a rest and good meals made the best possible prescription. But the season was growing later, and they had to reach Melville Bay before ice and more ice doubled its already impassable nature. Besides, there was no certainty that the sea would thaw enough to permit further passage. In time the lummes would fly away, and their island rookery would become a desert. Dog meat might postpone, but would never defeat, what would come next.

After a week on the island the ice still showed no sign of becoming passable, so the men continued fetching eggs and cooking lummes, proclaiming it the sweetest fowl known. Kane made his way to the top of a hill, one high enough to give him a view of Greenland's interior, frozen ocean. He described it as "a vast undulating plain of purple-tinted ice, studded with islands, and absolutely gemming the horizon with the varied glitter of sun-topped crystal." Never again would geologists insist that ice could not possibly be the carrier of Switzerland's largest erratic boulders.

Looking in the other direction, Kane could see Baffin Bay and the party's escape route to the south, still frozen over and as impassable as the ice sheet itself. The men stayed on the island for two weeks, regaining their strength, doing what they could to make the boats seaworthy. At last breaks in the ice seemed real enough to give hope of progress, and the party managed to round up the dogs and set forth once more. There was an immediate misfortune as one of the boats spilled over, tossing away the group's best shotgun and the kettle used for turning iceberg water into pretend tea. The party kept rowing, moving southward toward Melville Bay.

NIAGARA FALLS, JUNE 17, 1848...

Agassiz and a team of seventeen others arrived in Niagara on the train from Buffalo, New York, and booked rooms with views overlooking the Falls. The June flow was the highest of the year, as the snow runoff swelled the river. The flow was so great that the Falls themselves were almost invisible, hidden behind a huge cloud of mist that rose from the base and spread over the water, like clouds obscuring a high peak.

The team had come to Niagara on a sudden impulse, killing time while they waited for a boat in Buffalo that would take them up the Great Lakes to Mackinac Island. They played the role of tourists more fully than that of naturalists and made no scientific contributions to understanding Niagara's geology. They did make use of one tourist convenience that had opened for business since Lyell's visit, the *Maid of the Mist* boat ride. The team took the boat right up into the teeth of the cascade, disappearing into the mist and getting a clear view of the arc of the falling water.

Then it was back to the train for Buffalo and the team's real work. They were traveling north to Lake Superior, where they would take canoes and examine the lake's Canadian shore. The team included nine students from Harvard (where Agassiz had landed a professorship), several of Agassiz's colleagues from his summers on the Aare glacier, a writer who would produce a narrative of the voyage while Agassiz wrote of the scientific observations, and a couple of European naturalists who had decided that, with the Continent in full revolution, 1848 was a bad year to be in Europe.

Agassiz had brought a blackboard and planned to give regular lectures about what they saw. On their first day out, in Albany, he told the group to notice the erratic material of all sizes. They had sighted erratics all day as they rode the train westward from Boston, and they could expect to see more and more of them as they moved northward.

In Buffalo the team dodged the pigs that ran freely through the city. At the harbor they boarded a steamboat that would carry them to Detroit before noon the next day. Breakfast time of the day after that found them entering Lake Huron. They had now left the commercial bustle of the United States behind them. Michigan was off to port side, but it was still very thinly settled forestland. Boomtowns like Cleveland and Detroit had thrived after the opening of the Erie Canal, but those were on the southern Great Lakes. Even today Lake Huron supports no great cities. The team saw Native American canoes from time to time. The weather had turned chilly as well and would not really warm up until they were back below Lake Huron. They found that, up there, they needed the warmth of a campfire nearly every night.

The next morning they reached Mackinac Island, only to discover they had just missed the weekly steamer to Sault Sainte Marie. Agassiz found a way to pass the time. The Native Americans on Mackinac were fishers, and the professor soon introduced himself. He bought some fish, partly for food and partly to see what the local fish were like.

Instead of waiting a week to continue on, the team hired a boat to carry them to the Sault. During the voyage high winds forced them to land on an island, where they were greeted by the local pioneer. He looked as wild as John the Baptist but proved to be an educated man who recognized Agassiz's name. He disappeared into his house and returned with a preserved specimen of Lake Huron fish, a garpike, which he insisted Agassiz accept as a gift.

Eventually the wind calmed down, and the team moved on. They now entered the country of endless insect bites.

Mosquitoes swarmed about them whenever they landed, which, of course, they did every time they needed to sleep or eat. Nor would this be a torment confined to the narrows between Lakes Huron and Superior. They would complain about insect swarms for the duration of their tour in the north, and once they got beyond Sault Sainte Marie, they encountered blackflies, pests that make mosquitoes seem like timid creatures. Small enough to pass through mosquito netting, blackflies bite with the tenacity of bulldogs.

In the Sault they met another geological party. James Hall, the same geologist who had escorted Lyell through the Niagara country, was conducting an extensive geological survey of the area with a young man fresh out of college, Josiah Dwight Whitney. Hall loaned the team an extra canoe. Whitney, whose name is memorialized on California's tallest mountain, was more amused than impressed by the Agassiz team. Agassiz fancied himself as going on a scientific exploration, in the manner of Humboldt's five years up the Amazon. Whitney, however, described their adventure as a "tour of scientific pleasure," and this phrase seems to get at the heart of Agassiz's summer vacation. They were not pressing into a terra incognita beyond old maps. Indeed, they obtained a good map of Lake Superior while they were in the Sault. The Canadian side of the lake had a string of trading forts operated by the Hudson's Bay Company, while John Jacob Astor's people traded for fur on the American shores. Steamboats moved in these waters, and companies were interested in mining the lakeshore's extensive mineral wealth. (Two decades later Agassiz's son, Alexander, headed a copper-mining company on the Michigan coast of Lake Superior.) What the team did bring that was new was Agassiz's glacier-accustomed eyes. Just as he had taught Buckland a new way to see the familiar Scottish scenery, Agassiz could look at Lake Superior and see details that had gone unremarked.

Lake Superior is a gigantic trough surrounded by high walls. The many rivers that feed it all flow from the north, and each one has to tumble over an escarpment to reach the lake. Superior's north shore is a land of many cascades. Agassiz noted that the lake's northern wall is rugged, and yet the whole of its surface is flat. There are no peaks anywhere along it. Agassiz was certain the whole area had been ground down under a huge glacier. He could see the scratches on the rocks, and the lake's beaches were covered with enough erratic debris to make sleeping uncomfortable.

Beyond Sault Sainte Marie, the team traveled in canoes paddled by local guides. The first part of the Canadian lakeshore is known as Gaulais Bay, and when they camped on one of its beaches, Agassiz showed the team a polished rock that he claimed as total vindication of the Ice Age theory. The granite rock was several hundred feet wide and had been cut to form a glossy, smooth surface. Scratches showed the direction of the glacier as it had crossed the boulder, while a second set of grooves traced the route of the glacier's retreat when it melted. The angle of the grooves showed they could not have been made by anything floating in the lake. The uniformity of the polish also proved it could not have been worn down by water. Water erodes separate parts of a stone at different rates, depending on the variation in a rock's hardness. But this boulder had been worn down by something untroubled by whatever hardness the rock presented.

The team continued onward from one river to another, until they all seemed to blend into a single river. Many of them even had the same names—Blackwater River, Whitewater River, and the Montreal River repeated themselves endlessly over the miles. Agassiz occasionally pointed out rocks the shape of which bore the unambiguous traces of a glacier. He looked for rocks with roundings and scratches, like those he had seen in the Jura, that ran against the grain of their stratification. Aurora borealis (the northern lights), which Kane mentioned only rarely in his journal, was visible almost nightly here, although the lakeshore is still below 50° latitude. The sky was lit by curtains of flashing light, like an enormous lightning storm without thunder. During severe magnetic events, the northern lights are visible well below the Arctic Circle, but it is surprising that Agassiz saw them nightly. Paris, France, is slightly to the north of where the team camped, and it almost never gets sight of the lights. Perhaps it was aurora borealis that distracted the team one night as they let their campfires get out of control and set the forest on fire. They eventually retreated to their canoes while the whole woods was ablaze. Presumably, though nobody stayed to check, the fire was halted at a nearby swamp.

They traveled west across the lake for a month, getting almost to the U.S.-Canadian border. The farthest settlement they reached was Fort William (now called Thunder Bay), where they found letters and newspapers waiting for them. The mail had been forwarded from the Sault and sent by steamer to a nearby copper mine. For anyone familiar with the difficulties of real exploration, this journey seems almost too touristic in the way civilization was always hovering nearby; but Agassiz and his glacier eye kept at work. After a week's exploration of the Thunder Bay area, the team turned and began to retrace their route to the Sault.

The lake's Thunder Bay region is especially dramatic as islands and peninsulas jut out from the shoreline to create a series of walls and bays. Thunder Cape has an especially striking line of naked cliffs that drop straight down. The eastern entrance to this bay region is now called Terrace Bay, for the obvious reason that it contains natural terraces. Three main terraces and several lesser ones rise above the lake. Agassiz lectured the group about them, explaining that the terraces were formed by materials deposited after glaciers had shaped the land. He could tell the terraces came later because they lay on top of grooved and rounded rocks. The glacier, he calculated, had been a mile thick. Evidence for this came from the fact that *laves* end at about the 5,000-foot line on New Hampshire's Mount Washington. He dismissed Lyell's argument that there were no mountains near enough to Lake Superior to provide a glacier, noting that steep slopes are not necessary for glacier movement. "In the Swiss glaciers the motion is often slowest on the steepest part of the slope, and some glaciers of 7° inclination move faster than others with a slope of 40°."

They were back in the Sault by mid-August, two months after their arrival at Niagara Falls. Agassiz had acquired a huge collection of specimens and seems to have lost none of it. He packed his material into four barrels and twelve (mostly large) boxes and shipped them to Boston. As the group steamed back through Lake Huron, Agassiz gave a final lecture. He held up a small boulder, saying it was "an epitome of all the rocks seen." He talked about its great age, having been formed completely without any fossil content. It had been found far from where such rock was formed, yet it bore no marks of water transportation. Instead, it was uniformly grooved and rounded in a way that paid no attention to the rock's different degrees of hardness. He concluded, "I have no doubt, from similarity of its appearance in this respect to the rocks of the present glaciers of Switzerland, that it has been firmly fixed in a heavy mass of ice and moved steadily forward in one direction, and thereby ground down."

Besides the team members, other passengers on the ship heard this lecture. A clergyman objected to it, reminding Agassiz that the whole world had been made all at once. Agassiz refused to concede the point. The shocked clergyman insisted Agassiz admit that the Bible had settled this issue, but Agassiz admitted no such thing. Having run out of arguments, the clergyman could only express his indignation, which he did loudly and at length. The team members, after two months in their own little bubble of learning, were now reminded of what they were doing with their radical geology—smacking square in the face all defenders of tradition.

EDINBURGH, SCOTLAND, NOVEMBER 30, 1851...
James Forbes began coughing up blood, hemorrhaging badly and, it appeared, fatally. In those days blood in the lungs from tuberculosis was all too familiar a symptom and promised a short future. Forbes was a confident Christian and prepared for his Last Judgment, although, as it turned out, he had seventeen more years to live; however, after this incident Forbes never set foot on another glacier and did no

more fieldwork. The period of his active rivalry with Agassiz had lasted ten years, and Agassiz would continue for the rest of his life to act as though he had some great grievance against Forbes. Glacier motion had been Forbes's chief interest. He developed a verbal description of glacier motion but not a mechanical explanation for how it moves, nor had he nurtured any deeper understanding of the Ice Age.

The blue bands that Forbes had noticed did not explain how glaciers move, nor are they even a universal feature of glaciers. Banded "ogives," as Forbes's bands are known technically, are now said to be the effect of passage through an "icefall." The icefall is a glacier's equivalent of a rapids; as glaciers descend particularly steep ground, they tend to crack open into many crevasses and to create rough areas known as "séracs." During the summer the icefalls erode and thin the glacier. At the same time the glacier on the icefall turns dark because of windblown dust that falls into the crevasses in winter. The ice thickens and the séracs fill with snow, so after one year, the glacier has formed two new bands—a thin, darkened strip and a thicker white one. Although Forbes had been mistaken in assuming that the bands are the engines of glacier motion, they can measure glacier speed. Each pair of light and dark bands shows how much of the glacier moved across the icefall during one year. If there are five band pairs between the icefall and the glacier's snout, we know it takes five years for the ice in the glacier to cross that stretch.

Forbes's most enduring contribution was the founding of a tradition of British glaciologists that continues to this day, despite the absence of any glaciers in the British Isles. Forbes was followed by John Tyndall, a physicist best remembered for having explained why the sky is blue, but who, in the 1850s, explored the Alpine glaciers and provided a clear explanation of the mechanisms behind their formation, size, and movement. Tyndall's colleague in glaciology was Thomas Huxley, soon to be famous as the most aggressive defender of Darwin's theory of evolution, but none of this progress in understanding glaciers resolved the basic question of whether or not there had been an Ice Age.

Perhaps it would have helped reduce many arguments against the Ice Age if glacier mechanics had been understood from the beginning of the debate, but then, perhaps if they had understood, Agassiz and Schimper would never have been interested in the idea. The basics of glaciers are simple: They are formed when the annual snowfall surpasses the annual summer melt; they grow until they spread to an area warm enough for the melt rate to match the snowfall rate; over time the snow in the glacier crystallizes into compressed ice, and then the crystalline solids flow naturally, just as water does. As the ice at the glacier's snout melts, the crystalline ice behind it flows toward the front, turning the glacier into a kind of conveyor belt that carries the ice and anything else on it or in it toward the glacier's nose.

None of the disputants in the Ice Age controversy understood these simple principles. Agassiz imagined a sudden refrigeration plucking a glacier, like an idea, out of thin air, but it could not have happened that way. Snow, not a temperature drop, produces glaciers. On the other hand, Charpentier's theory (and Forbes's too) does not work either. They noticed that taller mountains tended to produce larger glaciers and concluded that the great Swiss glacier had been the result of higher mountains. They missed the point that melting increases as the glacier reaches higher temperatures; pulling the glacier farther up a taller peak is not going to sends its arm farther down into the valley. Sudden refrigeration will not pop a glacier across empty terrain, but you cannot grow one without a decline in temperature either.

The high-mountain theory of glacier size had another result. By blinding people to the matter of the melt rate, it made them doubt that glaciers could form on lowlands. The most scientific-sounding argument against British and American glaciers had been that there were no mountains high enough to create them. The fault in this reasoning is clear to anybody familiar with the nightmare that trapped Kane in a world of short summers. Even twenty-four-hour days could not melt away the ice before winter returned. No great mountains were needed, only persistent cold. The insistence that glaciers required lofty peaks was much like the pre–Wright brothers notion that a flying machine would need to flap its wings.

All of these subtleties were being worked out from a standing start of near-perfect ignorance, but nothing had happened to reduce human pettiness. Forbes was an intelligent man, but there seems to have been nothing noble about him. In his study of Swiss glaciers he was quickly persuaded that Agassiz was right. A great sheet of ice had once covered Switzerland; but Forbes did not immediately expand that idea to a belief in a general Ice Age. In 1845, however, Forbes changed his mind. Normally that year, he would have gone to Switzerland for the summer, but his health was already poor, so instead he traveled north to the Isle of Skye, off the western coast of Scotland. He had been there before (in

1836), but this time he was struck by the ample evidence of past glaciers. The hillsides had been scraped by glaciers, and stray blocks had been deposited in a manner "quite undistinguishable from those which a glacier would deposit under similar circumstances."

When Forbes published his observations, Sir George Mackenzie printed a response in the Royal Society of Edinburgh's journal denying that the glacier could have been the product of a much colder climate. Instead, he proposed that the hills on the Isle of Skye must have been much taller. Mackenzie was a bit of an oaf and had been laughed at when he tried to refute Agassiz's and Buckland's Scottish observations by saying that all the striations and erratics they reported had been carried in by icebergs during the biblical deluge. This new explanation hardly seemed any better. The top few thousand feet of a mountain are unlikely to erode away into oblivion while the traces of the glacier that the mountain supported still stand sharply on the surface. But, five years after Agassiz's Scottish tour, the sides had hardened, and Mackenzie's explanation for the Skye glaciation seemed preferable to the idea of a Scottish Ice Age. Forbes had no personal reason to defend the Ice Age theory and did not detour into Agassiz's side.

The next spring Forbes went to England's Lake District for a holiday with his wife. Buckland had advised him where to find glacier remains, and Forbes's notebook shows that he did record the presence of moraines, erratics, and *fresh* striations. Here was Forbes's chance to speak up. By then his authority in Britain was very strong as the Scotsman who had challenged the Swiss Agassiz on his home ground. The Lake District in northwest England lies nearly 200 miles south of Skye, and its hills are as much as 1,000 feet lower. The once-these-hills-were-taller explanation would have been an even tougher horse to ride here, but Forbes kept his observations to himself. He published nothing and lectured nothing. In the years before his hemorrhaging and breakdown in 1851, Forbes traveled extensively in Britain, especially in his native Scotland, and he saw glacier traces all around him. He again kept his mouth closed. Of course, it was not Forbes's job to prove the Ice Age in Britain. Agassiz had treated him badly; so let Agassiz stew. It was all very human and understandable but also petty and spiteful. History has the feel of inevitability to it, and heroics like Kane's struggle to bring his party home safely can seem superfluous until you remember what human littleness does. It delayed the congratulations and plaudits Agassiz wanted as discoverer of the Ice Age, and it kept Forbes in historical obscurity. As his health faded and travel to Alpine glaciers became more difficult, the countryside around him offered Forbes a great new subject to study. He saw the opportunity to become *the* expert on the Ice Age in Britain, and he passed it by.

THE NORTH WATER OF BAFFIN BAY,
JULY 21, 1855...

Fog and ice surrounded Kane's whaleboats. After sixty-six days of retreat, the party recognized Cape York looming through the mist. They had reached the entrance to Melville Bay and all its dangers. Kane aimed for shore, where the men could use their legs for something besides sitting and the dogs could scamper about without being nuisances. Kane informed his officers of his decision to break with their coastal course and head for the open sea. The change was natural enough. Two years before Kane had preferred to sail north through the open water, and, with this year's late summer, the wall of glaciers alongside Melville Bay was certain to make coastal travel impossible.

A voyage across open water demanded seaworthy boats, and the party hauled all three of its whaling launches ashore for examination and repairs. The *Red Eric* did not look like it could make the whole voyage, so its cargo was divided between the other two boats, and it was stripped of its furnishings as well. When the time would come to abandon it, the plan was that the party could use the boat for firewood.

There was one other ceremony to be performed. Kane wrote out a brief description of the party's desperate condition and intention of sailing across Melville Bay. He then deposited the message in a large cairn the men had erected. From the baggage he commandeered somebody's red flannel shirt and attached it to the cairn, with the thought that an alert sailor might see it and, if the party failed to reach the bay's southern end, obtain news of the expedition's fate. Having, in effect, raised their own tombstone, the men pushed off in their little boats for the open sea.

Their course was south by west, which took them out into broken ice. They were sculling into one of the earth's wonder-ways. Granted, it was cold—especially for late July. The gray sky and wind coming across the ice created the kind of sensation that, back in Kane's Philadelphia, might have made for a decent February day. The party used fox skins to keep from freezing to death, but all around were magnificent seascapes of ice mountains half hidden behind mists, and when the fog lifted, distant ice blinked on the horizon like twinkling stars.

Floes and bergs forced the party through a twisting course, as though it were following a corkscrew river. The sun never set; it just circled like a bird through the sky, sometimes moving east to west, other times swinging west back to east. Not surprisingly then, Kane woke one time to discover the party was lost. The helmsman had gone east around a berg when he should have gone west, and then the boats had followed the narrowing spaces between the floes. By the time Kane awoke, his fleet was trapped in a skinny little channel not wide enough to hold all three dinghies abreast. Worse yet, the zigzag behind them was closing and freezing, so they could not turn around.

The men were so disoriented that they still had not realized they were going the wrong way, and Kane kept the secret for a while longer. He said simply that they should dry out their clothes and things (which was true enough), and he had the boats hauled onto an ice floe. Taking McGary with him, Kane made his way over the ice to a berg, and, just as they had done the year before when they tried their escape to Beechy Island, the pair of them slipped and struggled their way up a 300-foot ice slope to survey their surroundings. Hardened whaler though he was, McGary burst into tears when he reached the iceberg's summit.

It was the same sight that had ended their previous failed escape attempt. The party was surrounded by ice and by Greenland's ice ocean. To the east rose the Greenland coast and the glaciers that launched their ceaseless armada of crushing bergs. A steady rumble of moving ice sounded faintly from that direction. To the north and south of them and to the west besides, there were icebergs and more icebergs, hundreds of them. And between the bergs lay a plain of frozen ice with no escape waters visible. The sight that set McGary weeping was no metaphor of an Ice Age. This was the Ice Age itself: shrunken in territory from its former glories, but just as ferocious and uninhabitable as ever.

If the rest of the crew had broken down at the news the way McGary did, the retreat might have ended right there in that frozen sea; and perhaps they would have broken down if Kane had still been an egotist trying to send men after his own glory. But Kane had become a leader, and the crew rose to the hour. They could not stay, they could not row, so they had to pull. The party set the boats on sleds and pulled them like a team of sled dogs. The *Red Eric*'s sled, however, had already been used for firewood, so they could drag only two of the boats. The *Red Eric* itself was taken apart and its planking added to the floors of the two remaining boats. It had taken only a few hours of sailing to reach their position, but it took three days for the exhausted, starving men to drag the boats back to the iceberg where they had gone wrong. Once back, they launched their two remaining boats and sailed west of the iceberg.

The dogs were not yet out of danger of being eaten. Kane calculated they had enough food to last them another eight days, but it could take more time than that to reach beyond Melville Bay. So they stayed within view of the floes and bergs in hopes of finding food. Only very occasionally did they see a bird. Sometimes they did manage to catch a lonesome bird, but one bird does not go far among fifteen starving men. Every once in a while the party spotted a seal, but those animals were too alert and fearful to let a whaling boat come into shooting range. Walruses were less cautious, but their tough hides meant the party had to get within harpoon range to catch one. One walrus did sleep so soundly that the party raised a lance, but suddenly the creature charged at the startled lancer and got away.

From time to time, the party rested on an iceberg, though that meant losing some hard-won distance because the berg was drifting slowly back to the north. Once, while resting, the *Faith* began to drift away, taking with it all the journals, charts, drawings, and logs that might justify the expedition. Without those documents, the men would no longer have any claim to the title of explorers. Even if they themselves survived, they would return as mere adventurers who had come through harrowing dangers sure to excite young boys, but without serving any larger ends. Besides, without the *Faith*, they would not survive. The boat carried half the party, and if it floated away, everyone would be squeezed into one lone craft of no more than 24-by-5½ feet. Screaming madly, McGary and Kane managed to hop onto a small ice floe and run to their boat, rescuing themselves, their documents, and their reputations in one tormented action.

Back at sea, the party grew increasingly exhausted and feverish. The combination of perpetual daylight, hunger, and frustrated hope for food made the men terrible sleepers, and none of the party ever got a true rest. A week after leaving Cape York, Kane decided these ice-floe hunts were not worth the time they cost. He aimed his boats farther west, well into open water and away from the ice. Their speed did improve, but now all their food was the thin ration they carried with them. Off to the east they could see icebergs drifting north, against the surface current. Other markers did not exist. The eastern wall of glaciers was too far to be heard or seen. The sun still made its buzzard circles around the whole sky. Kane used a compass to find their course. The men were starving to death on a diet of bread dust and tallow. The pangs of hunger had long since passed them by, but strength had gone as well. Rowing to effect was increasingly difficult. The morning and evening prayers they offered throughout the retreat marked the steady ebbing of strength, and the boats could no longer stay afloat by themselves. The men had to bail around the clock to keep from sinking.

Kane recognized the giant size of another walrus on a small ice raft 500 or 600 yards away. Impossible though the capture was, he pointed out the target and aimed the boats for it. All commands now were hand signals, to let the animal sleep. The men placed socks over the oars to muffle their sound. Kane saw now that it was not a walrus, but a great seal. They only had to get close enough to shoot it, although with the watchfulness of seals, that was like trying to get close enough to salt a robin's tail.

At 300 yards Kane signaled for the men to pull in their oars. Just one muffled paddle kept them moving. They had not quite yet reached shooting range when the seal suddenly sat up on his little patch of ice. He wanted to live too. His starving hunters still made no sound, but Kane saw despair cross every face. Slowly the seal moved to the edge of the ice and coiled itself the way seals do to dive to safety. Petersen fired his rifle in desperation, and without any wasted motion, the seal slumped dead onto the ice.

The silence ended then. The men began screaming and rowing with all the effort their hunger could muster. Kane, as hungry and desperate as the rest of them, wanted to order another rifle shot, to be certain that the seal would not stir itself and splash into the sea. He feared, however, that the shot would simply increase the hysteria of his men as they realized the seal was not yet theirs, so he stilled his tongue.

As soon as the boats reached the ice, the men tore at the seal with their knives and stuffed themselves on raw blubber. All the men had seen starving Eskimos gorge themselves this way on a sudden harvest of food, and Kane had sneered at Eskimo indifference to tomorrow. He did not sneer now as he joined in the gorging. Stuffing themselves at once seemed the surest way to win any sort of tomorrow. Once they were fed, some of the men started to laugh as people sometimes will when they have survived a close call. "The dogs are safe," was the joke of the hour. It was true, but just as true was the fact that the men were safe as well. They were going to finish the retreat.

They had entered seal waters and, within a day or two, had shot a second one. On August 1, after a voyage of ten days across the uncrossable Melville Bay, the party spotted Devil's Thumb at the bay's southern end. A few miles beyond was Wilcox Point and the glacier that Kane had once thought among the most impressive in the world.

The men saw an Eskimo in a kayak, but he ran from the sight of these gaunt, bearded ghosts in whaling boats. Their retreat only ended eighty-four days after it had begun, when the party was picked up by a boat carrying whale oil to the villages north of Upernavik. They heard their first news of the world in two years. War had broken out in the Crimea, but no one was quite sure where the Crimea was. More comprehensible was the news that the remains of the Franklin expedition had been found 1,000 miles west of Kane's Greenland camp.

Formally, the death of the Franklin party five years earlier meant the whole Kane voyage had been in vain and unnecessary, but the news the Kane expedition brought with it proved enduring. While Kane's party was making its retreat past the Greenland glaciers, the eighth edition of the *Encyclopaedia Britannica* appeared. Its entry on glaciers had been written by Agassiz's rival, James Forbes, and surveyed valley glaciers all over the world. Forbes discussed glaciers in Switzerland, the Himalayas, Norway, Spitsbergen, and Iceland, but of Greenland's ice-sheet glaciers he said only, "The western coasts of Greenland appear [*sic!*] to offer the same phenomena [the formation of icebergs], but on a grander scale." The facts of an ice sheet, a 60-mile-wide glacier, and a 1,000-foot glacier face were familiar to the scurvy-ridden survivors who had just come down to Upernavik and to the Eskimos they had left behind, but not to geologists. The Kane party had chased the Ice Age to its lair. The crew's documents and drawings would transform talk of an ice-shrouded world from an imaginary *maybe* into a palpable *is.*

But how to get the party and its news beyond Upernavik? For a time it looked as though the route home might lead through England. A Danish brig was headed that way, but at the last moment an American steamship appeared. It had come on a rescue mission to find Kane and his men. For one last time the party boarded the *Faith,* and the men rowed out to meet the newcomers. As they made their way aboard the steamship, the captain thought he recognized someone and asked, "Is that Dr. Kane?"

"Yes," came the answer, and cheers broke out from men in the ship's rigging.

THRUST HOME

BOSTON, MAY 28, 1857...

Time had caught up with Agassiz, and with Lyell too. The necessary poet had emerged. After Kane's return anybody could see great ice with Agassiz's eyes. Twenty years had passed since Agassiz had scandalized science with his theory of an Ice Age. Time and again he had pointed to evidence of huge glacier effects only to lose his learned audiences, who could not believe that his vision was possible. Now, however, Kane had described the Humboldt glacier to reporters from the penny press. His words had filled everybody's imagination with images of Melville Bay launching a wall of bergs into the sea. Great ice had finally become part of the world's stock images. Agassiz agreed to provide Kane with a blurb that was used to promote his two-volume, illustrated account of the years in Greenland. Agassiz was now the man who had understood ice all along. It was the perfect hour for him to stand up and say, *Now you can see what I meant when, in Edinburgh, I compared Ice Age Scotland to Greenland.*

But time had been passing for Agassiz. May 28 was his birthday, and the 1857 birthday was his fiftieth. He had only just turned twenty-four when he had sat with Humboldt to hear Cuvier's last lecture. Now Agassiz himself was one of the world's most admired naturalists, and the cream of Boston's learned society came out to honor him. Agassiz's club, called the Saturday Club because it met on the last Saturday of each month, held a special Thursday dinner in Agassiz's honor.

The guests, who came in sopping wet from a day-long rain, included the writers Ralph Waldo Emerson, Oliver Wendell Holmes, and James Russell Lowell; the creator of pragmatic philosophy, Benjamin Peirce; the historian John Lothrop Motley, whose *Rise of the Dutch Republic* is still worth reading; and the essayist, E. P. Whipple, who was famed for his combination of wit with high ethical concerns. Agassiz's coauthor of his book about Lake Superior, John Elliot Cabot, also came. Some able businessmen were there as well, while the master of ceremonies was the poet Henry Wadsworth Longfellow.

The festivities began at 3:30 in the afternoon and ran well past sunset, to 9:00 P.M. The guests feasted, smoked cigars, joked, and, of course, there was much speechifying about Agassiz. Longfellow, Holmes, and Lowell each read poems in Agassiz's honor. Longfellow's poem was the sort of thing that probably amuses when read aloud to a dozen friends who are murmuring "here, here" between cigar puffs, but it looks pretty flat on the cold page:

It was fifty years ago
In the pleasant month of May,
In the beautiful Pays de Vaud,
A child in its cradle lay.

And Nature, the old nurse, took
The child upon her knee,
Saying: "Here is a story-book
Thy Father has written for thee."

"Come, wander with me," she said,
"Into regions yet untrod;
And read what is still unread
In the manuscripts of God."

And the rest was even worse. Emerson must have been in a good mood at the feast, for he recollected that the poems were "all excellent in their way."

Emerson himself was the most famous speaker at the table. His lectures were in constant demand, but as an impromptu orator of the sort that Agassiz's fiftieth called for, he had no skills. Cabot said Emerson "rarely attempted the smallest speech impromptu, and never with success." There is no record of Emerson having said a word at Agassiz's fête, and he probably gave no speech; however, Emerson had become one of Agassiz's good friends. He appreciated Agassiz's ideas and saw him as a philosophical ally, jotting in his notebook, "Our thesis is that nature alone interests man, as man seems the object of nature.... Agassiz has expanded this & confirmed it. Nature is for man not man for nature."

That general appreciation was what had brought this roomful of people together, but what must strike any historian of science is how the greats who had assembled to honor Agassiz were humanists and literary figures, not other scientists. Boston did have other scientists who could have been invited. Elisha Kane's old geology teacher, William Barton Rogers, was now in Boston and was about to found the Massachusetts Institute of Technology. Asa Gray, botanist at Harvard, was one of Agassiz's colleagues. James Dana, in New Haven, was a geologist and enthusiastic supporter of the Ice Age theory. These fellow scientists and teachers might have seemed like Agassiz's natural companions and guests for a "golden" birthday, but they were like Lyell and Edward Forbes—willing to disagree with Agassiz to his face—and Agassiz did not much like that. Even as a young man he had been happy to be the lone genius presiding over a society of capable amateurs.

Agassiz's learning, enthusiasm, and eagerness to talk made him popular with philosophers and students. He had been in America for ten years and was having a marked impact on scientific education. His estranged wife, Cécile, had, most conveniently, died, and Agassiz had taken himself an American wife from an old Boston family. Henry Adams, writing about himself in the third person remembered his college years: "the only teaching that appealed to his imagination was a course of lectures by Louis Agassiz on the Glacial Period and Paleontology, which had more influence on his curiosity than the rest of his college instruction put together." Adams went on to be one more literary humanist, but Agassiz also taught many fine science students. William James concluded at the end of the nineteenth century, "There is hardly one now of the American naturalists of my generation who Agassiz did not train." Yet of new work in science, there had been little. Agassiz still taught the Ice Age, but he no longer tried to prove its reality.

After the trek to Lake Superior Agassiz went on no further expeditions to amass data about the Ice Age. He had not written either of his proposed second and third volumes of his *Système glaciaire*, the very volumes that were supposed to lay out the case for the Ice Age; nor had he done anything, even in the privacy of his own notes, to bring his Ice Age theory up-to-date.

Agassiz remained confident that his basic premise of an Ice Age was correct. His Scottish tour, along with all he had seen in New England and around the Great Lakes, was ample proof, but there were still problems with his original idea. It had become clear that the Ice Age had not erased all life, and it was also evident that the frozen world had held more than a single glacier. Agassiz had seen for himself that the groove markings in Lake Superior pointed in a different direction from those around Boston. Both areas had been covered by enormous glaciers but not by the same glacier.

These facts were not problems for the Ice Age theory, but they did pose difficulties for the meaning that Agassiz had assigned the Ice Age. They did not support Agassiz's brutal theology of a Creator sweeping away the old production in an instant, freezing mammoths in their tracks while spreading a glacier over the world from the North Pole to the Mediterranean Sea in Europe and to Long Island Sound in North America. But surely, if Agassiz had wrestled with the evidence, he could have revised his old idea without tossing out his faith. After all, even in the Noah story, some creatures from before the flood survived to live after it; but Agassiz had not reconsidered. Instead, he had quit gathering evidence about the Ice Age in America, and he had become more insistent than ever that there were no varieties, only separate species. This quarrel about varieties had become so well known that Emerson and Henry David Thoreau even joked about it between themselves, although they probably had no notion of Agassiz's dredging expedition with Edward Forbes or of how much of the meaning of the Ice Age was riding on this technical quarrel.

Emerson did know his friend well enough to have seen how his theology interfered with his science, and vice versa. He jotted in a notebook:

> *You must draw your rule from the genius of that which you do, & not from by-ends. Don't make a novel to establish a principle of polit[ical] economy. You will spoil both. Do not set out to make your school of design lucrative to the pupils: you will fail in the art & in the profit. Don't set out to please,—you will displease. Don't set out to teach theism from your Nat. History, like Paley & Agassiz. You spoil both.*

Perhaps sensing the limits of his approach, Agassiz urged other, more literary writers to take up his themes. Every author knows the discomfort of having others suggest subjects, but it is surprising to find Agassiz, who was himself a successful man of letters, prodding Longfellow to write a poem about what science had revealed of natural history. He described the successive eras starting with the primeval rock, proceeding to great fern forests, fantastic beasts, and so on. "There ought to be an epic about it," Agassiz told Longfellow. This moment is always an awkward one for writers, but Longfellow handled it coolly. He told Agassiz he had no doubt there ought to be such an epic, and there might be one, but he was not the man to do it.

Agassiz had been fiddling while time burned. He had arrived in America very much as a man of the future, with ideas and images ahead of his time, but time had come even with him now. The years of stalemate had ended. Kane's return from Greenland brought a change in the public imagination. Nobody quite yet realized it, for poetry does not send people running through the streets shouting, "Eureka." Rather, it changes the way people think about things. Soon enough people would begin to notice that the Ice Age no longer seemed like such a strange idea. They would discover that somehow they could imagine a world of glaciers. The scandal of the Neuchâtel Discourse and the rejections of the Scottish findings would become increasingly hard to understand. Kane's return was the moment Agassiz should have been waiting for. Now was the time to expressly link Kane's description to his own theory of a global freezing, but the Agassiz steam engine had grown used to his staleness, and he did not seize the day.

ZÜRICH, SWITZERLAND, AUGUST 16, 1857...
For the first time in his career, Lyell was plainly bringing up geology's rear, and he was in a hurry to recover. He awoke early and prepared for another day's travels. Many ideas and memories had been stirred during the past few weeks as he returned to the Alps for the first time in twenty years. Even in the cities his thoughts whirled in excited dances. In Zürich his reading had reminded him of old sights and new questions. The Ice Age could no longer be dismissed as "not plausible."

Great ice was one of the "actual causes" of geological change. Everybody within the blast range of the penny press suddenly knew that ice was pouring out on all sides of Greenland. Ice had not been in Lyell's original book, but it was real. It was also a catastrophe—an actual cause associated with sudden and violent changes in the Alpine terrain. So Lyell had come to Switzerland, he later confided, with the secret purpose of seeing "the proofs with my own eyes" that glaciers are "adequate to do all."

Having at long last decided to reconsider his Niagara apostasy sixteen years earlier, Lyell threw himself into his travels with the legs of a medieval pilgrim. He walked and climbed everywhere. His guides may have been used to leisurely tourists, but Lyell turned them into field geologists.

He forced one set of guides to spend hours with him while they searched near an active glacier to find a great erratic that had unquestionably been transported there by the glacier itself. And after hours they found one, a stone of many tons sitting alone on a mountainside, near a glacier, but far from its granite source. With another group of guides, Lyell went straight to the face of a retreating glacier and got down on all fours to wipe away the frost and dirt beneath the glacier wall. There they saw rock that had been under the glacier just six months earlier, and it was polished flat with straight, parallel grooves clawed into it.

On the morning of August 16, Lyell luxuriated in the modern convenience of a railroad car. Then he switched to a coach as he traveled through the Aare valley toward Agassiz's glacier. To the west of the valley rose the Jura. This whole region, in Agassiz's theory, had been under a glacier. At Solothurn Lyell descended from the coach and walked to the city's limestone quarries. The place yielded a limestone of almost marble beauty, and Lyell arrived in time to see an extensive limestone shelf that had just been exposed by the quarry's laborers. It had been covered by 8 feet of mud, which the quarry workers took to be glacial mud because they could see it was an unstratified mix of dirt and erratics of every size. Meanwhile, the exposed limestone's surface resembled the great flat boulder Agassiz had found in Gaulais Bay. The Solothurn limestone had been polished to a sparkling shine and was scarred with the parallel furrows that, in Canada, Lyell said were iceberg scratchings. At Solothurn, Lyell reported, everybody attributed those markings to "the doings of Charpentier's great glacier." And now Lyell too believed that a glacier had "walked across the great valley of Switzerland" from Mont Blanc and Monte Rosa to the Jura "with a thickness of some 4,000 feet," where it polished and scratched the shelf just uncovered at the quarry.

That night Lyell began a remarkable series of letters to his father-in-law in which he privately reported all the reasons for his renewed acceptance of what Agassiz had proclaimed twenty summers before. Lyell had almost erupted the previous night in a letter when he suddenly interrupted a discussion of Madeira botany with a remark that "the glacial phenomena appear to me more than ever wonderful," but then he got a grip on his secret and jumped at once to another topic: "There is much lecturing here [Switzerland] in small places, very like the United States." But after seeing Solothurn's polished limestone, he could contain himself no longer and blurted out the news of his change.

Lyell was now talking like a man who, having broken under torture, cannot stop himself from going on endlessly about something he has long refused even to mention. For the most part, Lyell's letters are familiar sorts of correspondence. When he was a young enthusiast they were rich in scenes and news, but as he matured the sights and thoughts in his letters were quick sketches. The ideas, especially, were more often alluded to than developed. He had other places for presenting his ideas in full, but the letters that began on August 16 made even the letters of his youth seem restrained and under control. Now Lyell went on for pages that must have amazed their recipient as he wrestled with the terrible fact that he, the modern, had been wrong, while Agassiz, disciple of the old-fashioned, had been right.

Lyell wrote how he had been rechecking the whole story of glaciers. Perhaps at first he thought there was a good chance the glacier theory would prove inadequate and his drift doctrine could prevail, but he found there were many voices against him. Agassiz's seeds had sprung all over the Continent, and Lyell learned that every important geologist in and around Switzerland believed in a great glacier. Geology's Irish monks were just waiting to pass on the knowledge they had guarded and studied for so many years.

In Heidelberg Robert Bunsen (remembered today for his laboratory burner) provided Lyell with a sound overview. In Neuchâtel Agassiz's longtime assistant, Edouard Desor, showed Lyell the erratics and scratchings of the Jura. Lyell talked with more of Agassiz's old colleagues, men like Bernard Studer and Arnold Escher von der Linth. All these men based their arguments on facts to be seen in the earth.

Desor took Lyell to the Jura's greatest erratic, the Pierre-à-bot. Lyell had seen it before, but now he let the sight work on him the way a critic takes in a painting. "Pierre à bot," he wrote his father-in-law, "is really more wonderful when seen again, than when I first beheld it, so vast and angular, so clearly resting on a limestone chain, with a great tertiary valley between it and the Alps."

In the valley between the lakes of Neuchâtel and Geneva, another guide pointed out spectacular evidence that a huge glacier had once filled the entire valley. High up on its walls were great sections that had been scratched and polished by something. Couldn't drifting icebergs have done this? No, said his guides. Besides the inadequacy of iceberg power, there was no evidence that the Alps had been underwater during any recent geological eras. There had been absolutely no marine fossils found in the moraines, gravel deposits, and glacial mud. This absence struck Lyell so forcibly that he mentioned it early in his first letter on glaciers, and then a few pages later in that same letter he wrote, "I ought before to have told you that after diligent search the geologists have been unable in any part of Switzerland to find a single marine shell in any moraine, or any part of the boulder clay." His forgetting what he had already said suggests the frenzy of confession that gripped Lyell as he wrote his letter far into the night.

Escher von der Linth—the same guide who had silently escorted von Buch through the Alps—pointed to sights that could not have come from water. He showed Lyell what Charpentier had discussed in 1834. Erratic boulders on the right side of the Aare valley came from one region, whereas those on the valley's left side came from another source altogether. Water flows would never preserve such a perfect separation, but solid ice would.

Lyell was suffering the agonies of a half convert who recognizes many truths in the new doctrine but wonders if he must abandon his most cherished old faith. He admitted to Escher that he was now drawn to the glacier idea, but he could not stand the argument that the Alps had been shaped rapidly by a violent catastrophe. Escher—still the diplomat—reassured Lyell that the Alps' great foldings had taken place very slowly. If people had lived there at the time, they would not have known that mountains were even being born. The glacier and its melting had been grand events with many sudden effects, but they had operated only in the context of an already established, ancient Alps.

With that testimony to his deepest principle, Lyell was won. He began seeing evidence of glaciers as support for his insistence on actual causes and unlimited time, and his letters began to include mocking references to von Buch and Élie de Beaumont, whose catastrophe-based theories did not include great ice. So many effects that Lyell saw in Switzerland matched what he had seen in Scotland—the way, for example, glacier debris blocks the entrance of small valleys into large valleys. With that recollection, Lyell renounced his old apostasy and wrote, "If the hypothesis now generally adopted here...be not all a dream, we must apply the same to Scotland, or to the parts of it I know best. All that I said [with Buckland and Agassiz] on the old glaciers of Forfarshire I must reaffirm."

With those words, Lyell transformed himself from a pro-drift reactionary to a great-ice modern, and he resumed his travels as an eager student of ice, not as a skeptic looking for proof of glacier inadequacy. He continued through Switzerland for several more weeks, examining ice effects and testifying to his conversion. Then he went into northern Italy, where he continued to see evidence of a superglacier, and he continued to instruct his father-in-law:

> August 29: *There is much brother hood among these Swiss. ...Real lovers of natural history and science.*
>
> September 10: *It is wonderful how abundant are those peculiar memorials of ice-action which neither the landslip, nor the avalanche, nor the vibrations of earthquake, nor any known agent but ice, can produce; and which the torrent, whether of mud or water cannot cause, but on the contrary immediately effaces.*

September 21: *A lofty mountain or ridge 2,000 feet high, called the Serra, running out of the great alluvial flat is nothing but the left lateral moraine of an ancient glacier. That it is so, I am now fully convinced.*

A 2,000-foot-high moraine implies, as Lyell well understood, a glacier of fantastic proportions. Even the Humboldt glacier, the snout of which had now become world famous, would have been dwarfed behind the Po valley moraine. The glaciers of present-day Switzerland had little in common with the great ice of the past, and Lyell now scorned Alpine ice as "pygmy glaciers." Indeed, Lyell noted that the markings left by ancient glaciers are even now much more impressive and abundant than the dwarfed actions of contemporary Swiss glaciers, and he criticized Agassiz for not having made enough of these "negative facts."

But Lyell knew that, despite whatever carping about this or that point he cared to raise, he had been wrong and Agassiz right, and eventually he would have to say so. The dramatic gesture would have been a telegram sent from Zürich to Boston—*Congratulations! It's all glaciers and no ice rafts here*—but Lyell was never that dramatic or self-effacing. Only in January, when he was back in London, did he write George Ticknor in Boston, knowing the news would immediately pass down the street to Agassiz, "My [summer] tour was unusually profitable, first in the glaciers and then the volcanoes. I came to the conclusion that Agassiz, Guyot, and others are right in attributing great extension to the Alpine glaciers. . . . I found Agassiz's map of the Zermat glaciers and moraines very correct and useful."

Part VI

Two Tragedies and a Triumph

LONDON, APRIL 12, 1858...

The meeting of the Royal Geographic Society came to order. The members in attendance were mostly stay-at-home savants and moneymen, but there were also one or two well-traveled gentlemen. Presiding over the meeting was Sir Roderick Murchison, the same Roderick Murchison who had challenged Louis Agassiz's glacier ideas at the 1840 meeting of the British Association in Glasgow and had then refused to accept any of his evidence that Scotland had been under a glacier. It would be still another four years before Murchison would tell Agassiz that he had become a believer in the Ice Age theory, but Murchison was already an enthusiastic supporter of Elisha Kent Kane, so he was unhappy when, during this meeting, Dr. Henry Rink, a Danish official who had spent several years in Greenland, rose to question Kane's account of the Humboldt glacier.

In some ways, of course, Murchison was quite used to this sort of thing. Victorian explorers were as quarrelsome as political sectarians. They loved to butt egos over who had found the higher mountain or whose spring was a river's true source, and usually Murchison could look on with

Olympian neutrality. With Kane, however, Murchison had quickly joined in the excitement over his return from north Greenland—and that had been a very high note of excitement indeed. Even before the rescue ship had landed, alert New Yorkers had spotted its approach through the Verrazano Narrows, with flags flying, and a mob had gathered on the sudden news to welcome the party back.

Newspapers had at once been full of the story—how the party had survived, had made a miraculous escape, how it might have seen the entry to an Open Polar Sea, and how the Humboldt glacier formed a "crystal bridge" between Greenland and North America. Kane made speeches about Greenland and then settled down to produce a two-volume account of his adventures. Even before the book appeared, the Royal Geographic Society presented, in May 1856, its rarely bestowed gold medal to Kane. Murchison organized a ceremony in London, where the medal was given to the American minister, George Dallas. Then Dallas—and, yes, this is the man for whom both Dallas, Texas, and Dallas, Oregon, were named—sent the medal to Philadelphia, where, at another public ceremony, John Crampton, the queen's minister to the United States, handed Kane his medal.

Kane said thank you and returned to writing, while Murchison began planning ways to lure Kane over to London. Kane was honored and enticed by the prospect of meeting the great British explorers of the Arctic, but he could not come before his book was done. Meanwhile, his publisher raised prepublication anticipation to a level that would impress even today's hype-savvy public. The book won endorsements from all manner of celebrity, including Agassiz. Astonishingly, once the book appeared, it proved worthy of the hubbub. Its many engravings provided striking images of the polar ice, and Kane's language put the wonder and power of glaciers into everybody's imagination. One diarist, George Templeton Strong, in New York City, wrote of Kane's book, "Franklin's [account of his] first journey has a strong tragic interest, but this far exceeds it in clearness and picturesqueness of description and conveys a much more distinct image of the perils and marvels of the polar ice." The volume became a kind of coffee-table book—"centre-table book" was the term Kane used for it—and people proudly exhibited it in their homes. It was said that for years afterward, a visitor to an ordinary American home could see only two books on display: the Bible and Elisha Kent Kane. Thanks to its many illustrations, even the nonreader learned from Kane what Greenland's conquering ice looked like.

Soon after the book appeared, Kane did depart for London, where Murchison had organized a grand festival in his honor. So Murchison was no mere presiding onlooker when Rink challenged Kane's work. Sir Roderick had his own reputation in this head-butting, and he defended Kane brusquely.

Rink, who appears never to have traveled much above Upernavik, made three points. He said that the Cape Forbes side of the Humboldt glacier was surely part of Greenland too, and not, as Kane supposed, on a separate island. Second, he doubted the accuracy of the reports that Kane had made about what lay beyond the Humboldt. Some people took that charge to mean that William Morton, the scout Kane had sent on to Cape Forbes, had never really gotten so far. Finally, Rink questioned Kane's description of the Humboldt as "the counterpart of the great river system of

Arctic Asia and America" that was "pouring out a mighty frozen torrent into unknown Arctic space."

The first two points would take decades to resolve. No other explorers would reach the Cape Forbes area until the 1880s. (And they would report that Morton's account was too accurate to have been an invention.) But Rink's argument that the Humboldt was not Greenland's equivalent of the Lena River, which drains Siberia, was a terrible distortion of what Kane had said. Rink insisted that many other glaciers besides the Humboldt were fed by the Greenland ice sheet, but Kane had said precisely that same thing. In the very passage that Rink challenged, Kane had said Greenland's interior ice moved "onwards like a great glacial river seeking outlets at every fjord and valley, rolling ice cataracts into the Atlantic and Greenland seas."

Why did Rink challenge Kane so unfairly? Priority! His challenge served notice on the audience that he had published the elementary facts of the ice sheet before Kane's book had appeared. Rink's report had seized no imaginations and did not lead to any reconsideration of Greenland's glaciers—James Forbes's essay on glaciers in the *Encyclopaedia Britannica,* for example, paid no attention to what Rink had said—but now that Rink had spoken up, everybody in the room could think, *Ah, yes, that is what he was talking about.*

Murchison immediately defended Kane and silenced Rink. This was the kind of fight for recognition that had kept Agassiz and Forbes battling for so many years. But in this instance, Kane never replied. He had fallen dead fourteen months earlier.

Kane had been physically ruined by his Greenland experience, and he did not recover at home. Like Ulysses S.

Grant, who raced to complete his memoirs while dying of throat cancer, Kane was a dying man as he pushed forward with his book. When he shipped out to London right after the book's publication, he was already desperately ill. He was too weak to stay long in Britain and left only eight days after his arrival. He missed the celebration scheduled in his honor and never did meet the many British explorers who had planned to come.

Kane was in Havana when he died on February 16, 1857, and his death produced a national demonstration of grief. His body was taken to New Orleans, placed on a riverboat, and carried up to Cincinnati before being loaded onto a train for Philadelphia. All along the way, at New Orleans and at the many landings on the Mississippi and Ohio Rivers, local orators addressed the mourning crowds about the heroic explorer and author.

Against that wall of dismay and the great image of the Greenland ice that Kane had placed in the public domain, a man like Rink could do nothing. He was a Danish bureaucrat who had spent a few years posted in the Greenland settlements and who then, in 1853, had published a brief report describing the Greenland ice sheet. It had been a good enough report, but it had not placed the Greenland ice into either the public or scientific imaginations.

But among scientists, who believe they progress without poetry, Rink's challenge to Kane had an impact. By speaking in London he reminded scientists that he held priority on the fact that Greenland's interior ice sheet is the source of its many glaciers and is the engine that makes Greenland's ice so much more monstrous than Switzerland's. From then on, when geologists cited Greenland's glaciers and its supporting ice sheet—and, from then on, they would cite them often—they would mention Rink as the source of the facts, rather than Kane as the source of their imagery. Elisha Kent Kane, celebrity-hero and poet-teacher, was already on his way to being forgotten.

LONDON, JANUARY 1863...

Greenland, Greenland. Lyell had just published a new book that was full of Greenland. Years earlier, when Lyell and Agassiz had spoken to the geologists in Edinburgh, nobody had understood as Agassiz compared Scotland to Greenland and Lyell spoke of the Antarctic. Now everybody knew about Kane and could follow the comparison. Lyell set the Antarctic aside. In his latest book it was all Greenland:

> *The argument...is now brought forward, and with no small effect, in favor of the doctrine of* ***continental ice on the Greenlandic scale.***
>
> *Norway and Sweden appear to have passed through all the successive phases of glaciation* ***which Greenland has experienced.***
>
> *In Norway and Sweden...the land is slowly rising; but we have reason to suppose that formerly, when it was* ***covered like Greenland with continental ice****, it sank at the rate of several feet a century.*
>
> *Some of these phenomena may now, as we have seen, be accounted for by assuming that there was once a crust of ice* ***resembling that now covering Greenland.***

> *The mountains of Scandinavia, Scotland, and North Wales has served, during the glacial period, as so many independent centers for the dispersion of erratic blocks,* ***just as at present the ice-covered continent of North Greenland is sending down ice*** *in all directions to the coast, and filling Baffin's Bay with floating bergs.*

More examples are available, but these make the point. Lyell was establishing Kane's imagery as a cliché of science. (Of course, he cited Rink, not Kane, but, to get away with it, he depended on what readers had learned from Kane, especially in that last sample.) In the future writers about the Ice Age—or "glacial period," as Lyell still called it—would point to modern Greenland as a way of grasping what the Ice Age had been like.

Lyell's new book, *The Geological Evidences of the Antiquity of Man,* was another success and required further printings in April and November. Like most of Lyell's books, this one was a compilation of facts and observations that others had made. Lyell organized mountains of data to make his points, overwhelming any opposition by the sheer volume of the evidence. In the course of this account, however, Lyell had to serve himself unusually large slices of crow pie as he repudiated many of his old positions. He had to swallow two old arguments in particular. Besides opposing the Ice Age, Lyell had, in all editions of his *Principles of Geology,* accepted the biblical idea that humans had been on earth for only a few thousand years. With this latest book, he now rejected that idea. At the end of the 1850s, British geologists had discovered a human skeleton mixed with the fossils of extinct and ancient animals. New and equally clear evidence then appeared in France, so that shortly after he accepted the Swiss glacier, Lyell also changed his mind about human origins. We were much older than previously admitted. In his new book Lyell made the case for his two great changes of mind; however, it would be naive to assume that Lyell was merely reporting the facts, as scientists sometimes do. James Forbes had not bothered to publish his turnaround, and although Lyell was a nobler man than Forbes, he was not merely groveling for science's sake. From his first publication to his last, Lyell was defending his great dogma that the world had always been shaped by the same forces and sizes of forces that change the world today.

Throughout his latest book Lyell showed how his old arguments could no longer stand. Of the Ice Age, he retracted many points. He had argued that the parallel roads of Glen Roy had come from a flooding when the sea had covered much of Scotland. Now Lyell accepted Agassiz's idea, specifically crediting him as its source, that a glacier from Ben Nevis had dammed up a lake in these side glens. Lyell had also said that Scandinavia's erratics had been floated in on icebergs. Now he admitted that an ice sheet had covered Scandinavia. Ireland, Switzerland, Scotland, Wales, Scandinavia, Belgium—they had all been under ice.

Lyell's admissions seemed an amazing turnaround, and yet somehow it was no turnaround at all. As he had shown in his revised editions of the *Principles,* Lyell was a master politician who conceded everything to his enemies and yet somehow advanced his own purposes while crushing theirs. Many reviewers at the time complained that Lyell's book had a big, boring digression at its center—the middle 163 pages (a third of the book) devoted to the glacial period.

The material struck many readers then and now as having nothing to do with the title matter of human antiquity, but Lyell's primary interest was still his lifelong dogma. By itself human antiquity had little relevance to Lyell's main interest, but in the context of an Ice Age it was a terrible weapon against his most bitter enemy, the truest of Cuvier's disciples, Louis Agassiz.

Buried in the mass of facts about how Agassiz had been right and Lyell wrong, Lyell made the following assertion: When northern Europe had been covered in ice, the fringes of the ice supported species that, before the ice sheet, had been found farther north. This was the argument that he and Edward Forbes had made with Agassiz just before the professor had sailed to America. Edward Forbes had good fossil evidence of the migration of species up and down the seas as the climate changed. Agassiz had been unwilling to concede even the survival of a mollusk species during his Ice Age, but Lyell was shrewd enough to know that he needed something more important than shellfish before his position could carry the day, something like humans. Human species migrated with the climate too. Lyell argued that while toolmaking people lived in France, "we might expect to find Scandinavia overwhelmed with glaciers, and the country uninhabitable by man."

This argument was Lyell's coup. It allowed him to accept Agassiz's Ice Age without accepting the Neuchâtel Discourse. The meaning of the Ice Age that had so excited Agassiz and Schimper almost thirty years before was now buried. The ice had not catastrophically wiped out an entire world, forcing a new creation. Pre–Ice Age species in the north had survived by the simple expedient of moving south. There had even been people, alive and making tools, while the ice sheet froze northern Europe, just as today we prosper while Greenland freezes.

Agassiz had been jujitsued, flipped on his back by the force of his own arguments. He had met a fate opposite from Kane's. In the scientific world Kane had no standing. Writers used his Greenland imagery and cited Rink as their source. Meanwhile, they cited Agassiz as the source for their Ice Age notions while using Lyell's meanings. In the coming years Lyell's book would be the guide for the new discussion of the Ice Age and human origins. Two great books would soon follow: John Lubbock's *Pre-Historic Times* (1865) coined the term "prehistory," while James Geike's *The Great Ice Age and Its Relation to the Antiquity of Man* (1874) clarified the link between human antiquity and the Ice Age. Both of these books were profoundly shaped by Lyell. There were plenty of weak spots in Lyell's book. He still had North America underwater and free of glaciers, and some of Europe too was still explained as drifts during an inundation. Lyell also suggested that the Swiss glacier may not have come at the same time as the northern glacier. Other writers would clear these points up, but, just as his *Principles* in 1830 had laid the ground rules for discussing geology, Lyell's *Antiquity of Man* was the model for the new geological archaeology. Agassiz's *Études sur les glaciers* and his Neuchâtel Discourse no longer mattered.

Ignorance had gone in, understanding had come out, and Louis Agassiz was not happy with the results. He had hit all the long balls, yet he lost the game.

Ice is a major geological force. Lyell had missed it; Agassiz proved it. Yet Lyell was the one with the championship grin.

Scotland had looked like Greenland. That unexpected image that had so baffled an Edinburgh audience of savants had come from Agassiz. Now everybody could picture it, but they could imagine it without accepting Agassiz's metaphysics as well.

The modern earth has only recently emerged from a catastrophe. Lyell hated that idea, had devoted his life to fighting it. Now he granted it, without granting that it had been as catastrophic as Agassiz had insisted.

Agassiz stirred bitterly. It was as though he had finally caught on to the disturbing implications that had so distressed his audiences a generation before. Lyell's Ice Age was brutal, murderous, and part of no larger plan. For Agassiz, the Ice Age had always been providential, God's deliberate intervention in history. The mass extinctions had been divine too, a step toward the creation of a whole new world. Despite his quarrels with traditional Christians like the clergyman on Lake Huron, Agassiz could be and was a regular churchgoer because he thought his science could find the real truth of Providence. Now he got to see what the Ice Age looked like when it was not at all providential. It was indifferently brutal, wiping out mammoths and other species for no greater purpose than a rising river has when it floods a rabbit hole.

Ever since his dredging expedition with Edward Forbes, Agassiz had been warned that this day might come. When a preglacial variety of shellfish spilled onto the deck of their boat, Agassiz saw the future. Seventeen more years passed before Lyell produced his Ice Age book, plenty of time for

Agassiz to think anew, but he had not reconsidered. Yet Agassiz was still a steam engine when he wanted to be, and now he tried to fight back in the one way he knew how. He organized a new expedition for facts. He led a party up the Amazon River in search of proof that the Ice Age had extended to the equator. That idea, if proven true, would end the Lyell–Edward Forbes migration theory. It was also a new notion for Agassiz. In the Neuchâtel Discourse he had said that the northern glacier had reached only to the Mediterranean. In his account of Lake Superior he had plainly argued that the North American glacier had stopped at about the thirty-eighth parallel. But that kind of limited Ice Age was no longer sufficient for his philosophic purposes.

Along with Lyell's sudden judo maneuver on the Ice Age, the theory of transformation of species was back. Darwin had given new power to the old idea that today's species are descended from yesterday's fossils. That theory would fail, however, if it could be proven that the Ice Age had severed all links between preglacial and postglacial species. In 1864, Agassiz gave a lecture at Boston's Lowell Institute in which he repeated his insistence that the Ice Age had destroyed all life on the planet, and he had urged an exploration of Brazil and the Andes to obtain evidence of glaciers on the equator. Agassiz was then in his middle fifties, but he could still get things done; and by the spring of 1865, he had assembled the money and the people needed for such an expedition. As with Lake Superior he organized a team of students and other scientists. His American wife came along too.

One of the Harvard students to join the trek was William James, the future star of American psychology. James remembered Agassiz as bringing all the excitement of a schoolboy to the Brazilian adventure. In the middle of the night, as they moved upriver, James would suddenly hear Agassiz calling him to say he was too excited to sleep. That was how it had been for Charpentier in Bex, for Schimper in Neuchâtel, and for Buckland in the wilds of Scotland. Each of those energetic men had caught hold of a comet and could only ride it. Agassiz was still in full pursuit of facts, still the penetrating observer who could bring a thousand previous observations and details into consideration of each new fact. Each time the team saw a strange plant, insect, bird, or fish, Agassiz could link it to the whole natural history of creation, but this time all of Agassiz's facts were being assembled to support a thesis that James thought a humbug, the same doctrine Agassiz had been pursuing his whole career. "The study of nature has one great object," was the way he phrased it in 1862 before a Brooklyn audience: "It is to trace the connection between all created beings, to discover, if possible, the plan according to which they have been created, and to search out their relation to the great Author."

When he returned to the United States, Agassiz immediately announced that the expedition had been successful; Providence had been restored a role in natural history. Agassiz reached New York on August 6, 1866. On August 12, in Washington, D.C., he read a paper titled "Traces of Glaciers Under the Tropics" to the National Academy of Sciences, saying that Brazil, both in the Amazon valley and up in the Andes, had been covered by a glacier and that no physical relationship was possible between pre- and post–Ice Age species.

Instead of a scandal, the paper produced embarrassment and scorn. The Academy of Sciences never published the paper. Lyell laughed at it. In one letter Lyell chuckled, "Agassiz...has gone wild about glaciers....The whole of the great [Amazon] valley, down to its mouth, was filled by ice. ...He does not pretend to have met with a single glaciated pebble or rock...and only two or three far-transported blocks, and those not glaciated."

Agassiz made no public reply, although he did scrawl his bitterness in pencil on a piece of paper:

> *Sir Charles Lyell makes very light of the explanation I have given of the facts I have observed at the mouth of the Amazon. He is quite welcome to his criticism. But what I insist upon and what he does not seem to understand is the fact that that immense river has no delta as the Mississippi, the Nile, the Ganges have....The Amazon on the contrary loses daily ground by the gradual encroachment of the ocean and it rolls its twisted waters over a bottom which is not of its own making....But the real Amazon bottom...is occupied by a widespread deposit of the finest laminated mud; strikingly resembling the mud deposits observed everywhere in the Alps, where glaciers terminate at the margin of an extensive flat, over which the muddy streams issuing from under the ice have a chance of dropping their load.*

Agassiz had seen something, but no geologist today believes that silt had anything to do with a glacier. As it happens, geologists do believe that 230 million years ago the land that is now the Amazon basin was covered by a glacier, but that was before today's oceans and continents appeared. Agassiz saw none of the evidence of that time.

Agassiz was not finished. In the coming years he found clear evidence that the Rocky Mountains had held extensive glaciers, and he showed that when North America sat under an ice sheet, the southern Andes had also been covered by one. These proofs were solid additions to the Ice Age story, but they were not the additions that Agassiz wanted. He was filling in the details of an event that no longer proved what he needed to prove.

In 1872, the year before he died, Agassiz added a post-script to a letter to Karl Gegenbaur, a German anatomist and evolutionist: "Don't you find a meaning in nature?" he asked. "Is not this world full of the most wonderful combinations—just think of man himself.... From whence comes intelligence, and how is it that we ourselves can stand in a connection with this world which we can understand?"

It was a deep question, one raised by his own life. He had seen humanity move from near-total ignorance of the conquering ice to recognition and understanding of an Ice Age. He as much as anyone should have been in a position to say how that understanding had come. Tragically, however, Agassiz had never recognized that the trail he scouted was not the one Cuvier had described. Agassiz still thought he had simply traced the logic of his facts, and he died (on December 13, 1873) without ever noticing that all the critical accomplishments needed to establish the Ice Age had ignored logic. His bull sessions with Schimper, Kane's Greenland imagery, and Lyell's seizure of the Ice Age constituency had nothing to do with the process Cuvier described in his last lecture. If Agassiz and the others had not defied reason, they could never have gotten beyond the ignorance of 1830.

When Agassiz died the world knew it had lost a hero. Boston's papers put black borders around their front pages. Harvard's elite, Boston's elite, and other distinguished mourners, including the vice president of the United States, attended the funeral. An erratic boulder weighing more than a ton was brought over from Switzerland and placed as a marker on the grave, the first of many Agassiz memorials. By now there is a multitude of schools, libraries, and streets named after him. Statues and plaques lie scattered around the globe in his honor. There is a Mount Agassiz in California and another along the Chile-Argentina border. Even a formation at the bottom of the Pacific Ocean bears his name, while a lake that existed at the end of the Ice Age, an expanse greater than Lake Superior, has been named Lake Agassiz. In 1882, the people of Neuchâtel held a ceremony commemorating the fiftieth anniversary of his arrival there, and, in 1896, Boston held a ceremony commemorating the fiftieth anniversary of his landing in America, plaques all around.

Rationalists may shake their heads. Agassiz, after all, was only human, and he made many errors—wrong about varieties, wrong about evolution, wrong about how glaciers moved, wrong again about the origin and effects of the Ice Age. His memorializers, however, appreciate the importance of seeing anew, and they understand that he did what only humans can. Agassiz, combined with Kane and Lyell, gave us a new way to understand our globe and our connection to its history.

In one of Agassiz's memorials a former student named Nathaniel Shaler recalled what it was like to study under the professor, and the recollection has become a much-told tale. Ezra Pound even began his book *The ABC of Reading* by citing this "parable." Agassiz's preferred method of teaching was to hand a student some object, typically a fish, and then leave him to study it. In this case Agassiz gave Shaler a small fish and told him to learn all he could by looking, but he was to discuss the fish with no one, nor should he read anything about it. Agassiz wanted to hear what the student could learn from observation alone.

From time to time, in an exercise that lasted several weeks, Agassiz would ask Shaler what he knew and then, after listening to the reply, say, "This is not right." What was there to see? The scales, of course, formed an intricate pattern. The gills, when examined closely, are more than slits. They have all sorts of details that, without reference to a book, have no names but that Shaler could examine and describe. He could explore the fins, tail, eyes, mouth. If you look carefully at a fish, you may notice nostrils where a fish nose would be. What do they suggest? How is the lateral line along the fish's middle formed? By taking a mental step back, Shaler could note the relation among organs. Which ones were in line? Did they match up on each side of the fish? Were all the fins constructed the same way? Eventually Shaler was able to make a profound enough report to satisfy the professor and astonish himself.

Pound stopped the story there. Seeing the truth afresh, he said, was the essence of modern imagination, though why limit it to "modern" he did not say. Columbus and every discoverer, modern or ancient, has left the labels behind to find reality, quivering and exposed. Mistaken labels are why stay-at-homes can always laugh at explorers, for all of them—Columbus, Kane, and Agassiz included—went looking for one thing and found another.

But there is a coda to Shaler's story, one that Pound omitted. After Shaler had carried himself beyond map reading, he was assigned further observation projects, and one day he found that his old lessons contained an error. Agassiz classified fish by their scales, but this student saw a fish that had one kind of scale on one side and another kind of scale on its other side. This discovery startled the student and made him wonder about his teacher's wisdom after all. Eventually he screwed up enough courage to point out the problem, and Agassiz replied, "My boy, there are now two of us who know that." He had taught another person the importance of seeing the world through open eyes.

Acknowledgments

Dr. Albert V. Carozzi, Department of Geology, University of Illinois, and translator of many of Louis Agassiz's major writings, provided encouragement and information, and then he corrected the manuscript's most outrageous errors. Any errors in this book are all mine, while several of the correct points are his.

Charles Cowing, chairman of the Elisha Kent Kane Historical Society (New York), generously showed me the society's collection of Kane memorabilia.

Dana Fisher at the Ernst Mayr Library, Museum of Comparative Zoology Archives, Harvard University, immediately steered me through the library's special collection of manuscripts and volumes related to Louis Agassiz. The library gave me permission to quote from its manuscripts.

Peter Friederici of Flagstaff, Arizona, translated Agassiz's German account of his actions with Schimper.

Scott DeHaven at the library of the American Philosophical Society (Philadelphia) was most helpful in locating manuscripts and hand-drawn maps related to Elisha Kent Kane. Quotations from the library's manuscripts appear with the society's gracious permission.

Martin J. Rudwick suggested further reading sources for me to explore.

I was also helped by some people on-line. Wendy R. Tordoff and S. J. Martens were the first two members of the sci.astronomy.amateur newsgroup to report just what Kane saw during the occultation of Saturn. Patricia Kennedy at the soc.history newsgroup provided me with a good consideration of what Agassiz's £21 award in 1834 was worth in today's money.

Notes

Part I: Ignorant, Ambitious Men

For a good general account of the quest for an Open Polar Sea, see Berton, Pierre. *The Arctic Grail: The Quest for the North West Passage and the North Pole, 1818-1909*. New York: Viking, 1988.

p. 6 Text of Maury's orders given, from Kane, Elisha Kent. *The U.S. Grinnell Expedition in Search of Sir John Franklin*. New York: Harper & Brothers, 1853, p. 492.

6 Report on Maury Channel, from: ibid., p. 499.

7 Kane's description of his plan, from: ibid., p. 544.

8–9 For a solid survey of Lyell's *Principles*, see: Rudwick, Martin J. S. Introduction to *The Principles of Geology*, by Charles Lyell. 2 vols. Chicago: University of Chicago Press, 1990.

9 Lyell's error on Baffin Bay can be seen in: ibid., vol. 1, p. 109.

10 "[Lyell] altered the whole tone of one's mind" Quoted in "Lyell, Sir Charles Baronet" located on *Britannica CD 97*. Chicago: Encycolpaedia Britannica, 1997.

10 For an introduction to Cuvier, see: Rudwick, Martin J. S. *Georges Cuvier, Fossil Bones, and Geologic Catastrophes: New Translations and Interpretations of the Primary Texts*. Chicago: University of Chicago Press, 1997.

10 The facts of Cuvier's final lecture come from: Appel, Toby A. *The Cuvier-Geoffroy Debate: French Biology in the Decades Before Darwin*. New York: Oxford University Press, 1987.

14 For Lyell's comments on the Humboldt-Cuvier rivalry, see: Lyell, Charles. *Life, Letters, and Journals*, edited by his sister-in-law Katherine M. Lyell. 2 vols. London: John Murray, 1881, vol. 2, p. 128.

16 There is no better account of Kane's adventures in northern Greenland than that found in Kane's own two-volume report: *Arctic Explorations in the Years 1853, '54, '55*. 2 vols. Philadelphia: Childs & Peterson, 1856.

19 For Kane's father on glory, see: Corner, George W. *Doctor Kane of the Arctic Seas*. Philadelphia: Temple University Press, 1972, p. 85.

19 Kane's comment about the captain being proud of his ignorance, from: unpublished manuscript letter in the library of the American Philosophical Society, Philadelphia, ca. 1851.

20–21 "Am I not injuring my dignity...," from: Corner, *Doctor Kane*, p. 117.

23 The encounter between Forbes and Venetz is reported in: Cunningham, Frank F. *James David Forbes, Pioneer Scottish Glaciologist*. Edinburgh: Scottish Academic Press, 1990, p. 24.

25 The chief source for details of Agassiz's life is: Lurie, Edward. *Louis Agassiz: A Life in Science*. Paperback ed., with additions. Baltimore: Johns Hopkins University Press, 1988.

28 Emerson's anecdote on Agassiz and money, from: Emerson, Ralph Waldo. *The Journals and Miscellaneous Notebooks of Ralph Waldo Emerson*, ed. Susan Sutton Smith and Harrison Hayford. Cambridge, Mass.: Bel-

knap Press of Harvard University Press, 1978, vol. 14, p. 173.

29 Agassiz's rejection of Lamarck is quoted in: Lurie, *Louis Agassiz*, p. 83.

Part II: Trading Ignorance for Action

34 Kane's irony on leadership is from: Corner, *Doctor Kane*, p. 128.

37 Lyell's first letter to Agassiz is housed at the Museum of Comparative Zoology, Harvard University, Cambridge, Mass. Author's translation.

44 Charpentier's first reaction to glacier theory, quoted in: Carozzi, Albert V. "Agassiz's Amazing Geological Speculation: The Ice Age." *Studies in Romanticism* 5, no. 2 (winter 1966): 59. The original, untranslated quotation is in Charpentier, Jean de. *Essai sur les glaciers et sur le terrain erratique do basin du Rhône*. Lausanne: Marc Ducloux, 1841, pp. 241–242.

46 Hayes describes Melville Bay in: Hayes, Isaac I. *An Arctic Boat Journey in the Autumn of 1854*. New ed., enlarged and illustrated. Boston: Ticknor & Fields, 1867, p. 194.

47 "Blank wall of glacier..." from: Kane, Arctic Exploration, vol. 1, p. 37.

47–48 Kane's comparison of Greenland with the Alps, from: Kane, *The U.S. Grinnell Expedition*, p. 446.

48 "...gemwork...carbuncles," from: Kane, Arctic Exploration, vol. 1, p. 37.

49 Lyell cites Charpentier and Venetz in: Wilson, Leonard G. *Charles Lyell: The Years to 1841: The Revolution in Geology*. New Haven, Conn.: Yale University Press, p. 497.

52–53 Descriptions of the area around Bex, such as that of the blocks of Monthey, come from James Forbes's splendid travel account of glacial Switzerland: *Travels Through the Alps of Savoy and Other Parts of the Pennine Chain, with Observations on the Phenomena of Glaciers.* Edinburgh: Adam & Charles Black, 1843. Other descriptions of the land come from: Louis Agassiz. *Studies on Glaciers Preceded by the Discourse of Neuchâtel*, ed. and trans. Albert V. Carozzi. New York: Hafner Publishing, 1967.

61 Kane's reference to Constantinople, from: Kane, *Arctic Explorations*, vol. 1, p., 64.

64–74 The crucial sequence between Schimper and Agassiz is very poorly documented. Agassiz's fullest account is in a polemic: Agassiz, Louis. *Erwiederung Auf Dr. Carl Schimper's Angriffe*. Neuchâtel: Privately printed, 1842. Translated for this project by Peter Friederici.

69 "impossible at present to outline their respective parts," Carozzi, "Agassiz's Amazing Geological Speculation," p. 63.

71 "Animals are only the persistent fetal stages of man," quoted in Lurie, *Louis Agassiz*, p. 28.

72 Goethe: "an epoch of great cold" quoted in Agassiz, *ibid.*

74–77 Lyell's conversation with Lord Holland is reported in Lyell, *Life, Letters, and Journals*, vol. 2, p. 8. For the Buckland book under discussion, see: Buckland, William. *Geology and Mineralogy Considered with Reference to Natural Theology*. 2 vols. Philadelphia: Carey, Lea & Blanchard, 1837.

79–80 Kane's diagnosis of the dogs' psychological state, from: Kane, *Arctic Explorations*, vol. 1, p. 156.

81–85 The complete text of the Neuchâtel Discourse is given in: Agassiz, *Studies on Glaciers*.

86 Von Buch's cry to Saint Saussure, from: Carozzi, "Agassiz's Amazing Geological Speculation," p. 66.

88 Charpentier's explanation of troubles, from: Charpentier, *Essai sur les glaciers*, p. ii.

90 Diarist on Jura hike: Hallam, Anthony. *Great Geological Controversies*. Oxford: Oxford University Press, 1983, p. 93.

90–93 Discussion of Lyell's travel on the Rhine and his letter to Darwin is found in Lyell, *Life, Letters, and Journals*, pp. 20–21.

Part III: Changes of Heart

92–98 "occupied through nearly its whole extent...," Kane, *Artic Exporations*, vol. 1, pp. 226–228.

97–98 Agassiz's letter to Jules Thurmann, from: Marcou, Jules. *Life, Letters, and Works of Louis Agassiz*. 2 vols. New York: Macmillan & Co.,1892, vol. 1, pp. 124–125.

101 "It looks as though winter must catch us," Kane, *Arctic Explorations*, vol. 1, pp. 226–228.

108 Buckland's account of the last catastrophe, from: Buckland, *Geology and Mineralogy*, vol. 1, pp. 80–81.

110 Bad news from Buckland's wife, from: Imbrie, John, and Katherine Palmer Imbrie. *Ice Ages: Solving the Mystery*. Short Hills, N.J.: Enslow Publishers, 1979, p. 36.

110–111 The attempt to escape before a second winter is fully described in: Hayes, *An Arctic Boat Journey*.

111 "I cannot disguise it...," from: Kane, *Arctic Explorations*, vol. 1, p. 349.

114 "A chapter has been introduced," Lyell, Charles. *Principles of Geology: Or, the Modern Changes of the Earth and Its Inhabitants, Considered as Illustrative of Geology*. 3 vols, 6th ed. London: John Murray, 1840, p. xi.

115 "It is well known...," from: Lyell, Charles. *Principles of Geology: Or, the Modern Changes of the Earth and Its*

	Inhabitants, Considered as Illustrative of Geology. 3 vols. 6th ed. London: John Murray, 1840, p. 384.
117	Agassiz's presentation to the British Association meeting in Glasgow is summarized in: "Tenth Meeting of the British Association for the Advancement of Science." *Athenaeum* 673 (September 19, 1840): 729–755.
119	Murchison's letter about the Glasgow meeting, from: Rupke, Nicolaas A. *The Great Chain of History: William Buckland and the English School of Geology (1814–1849)*. Oxford: Clarendon Press, 1983, p. 100.
119–124	The travels with Buckland have been worked out in the most detail by: Cunningham, *James David Forbes*, pp. 54–56.
121	"One of the grandest natural phenomena...," from: Lyell, *Life, Letters, and Journals*, vol. 1, p. 158.
124	"Lyell has accepted your theory *in toto*" Rupke, *The Great Chain of History*, p. 100.
128–129	Maclaren cites Agassiz in: Cunningham, *James David Forbes*, p. 59.
129	Scotland declared irrelevant to Agassiz's theory, see: ibid., p. 29.
129	Kane sees sun going south again, from: Kane, *Arctic Explorations*, vol. 1, p. 405.
130–131	Text of Kane's marker: ibid., vol. 1, pp. 346–347.
133	Kane begins to express self criticism: ibid., vol. 1, p. 438.
133–134	Kane's dread of mysticism, from: Berton, *The Arctic Grail*, p. 288.
135	"What can they hope for...," from: Kane, *Arctic Explorations*, vol. 2, p. 108.
135	Kane cites Milton and Dante in: ibid., vol. 2, p. 57.
136	"Alas for Hans...," from: ibid., vol. 2, p. 235.

Part IV: Rock Bottom

For Henry Adams, see: Adams, Henry. *The Education of Henry Adams*. New York: Library of America, 1990.

Lyell's visit to North America is detailed in: Lyell, Charles. *Travels in North America, with Geological Observations in the United States, Canada, and Nova Scotia*. 2 vols. London: John Murray, 1845.

For a good account of Niagara Falls, see: Berton, Pierre. *Niagara: A History of the Falls*. New York: Penguin Books, 1998; reprint, Canada: McClelland & Stewart, 1992.

142 "...fairyland...," from: Lyell, *Life, Letters, and Journals*, vol. 2, p. 61.

144 Lyell's account of North American drift, from: Lyell, *Travels in North America*, p. 49.

145 Lyell on recency of drift: ibid., p. 52.

148 "...plausibility...," from: ibid., p. 136.

149 Lyell denies there was a dam to support water: ibid., pp. 106–107.

149–152 The Forbes-Agassiz story is most fully told in Cunningham, *James David Forbes*. Forbes himself assembled a series on the issue: *Occasional Papers on the Theory of Glaciers, Now First Collected and Chronologically Arranged*. Edinburgh: Adam & Charles Black, 1859. For Agassiz's opening salvo, see: *Monsieur Agassiz, Vous apprécierez les motifs qui m'ont engagé*. Public letter, dated April 11, 1842. Neuchâtel: Privately published, 1842.

150 Forbes on his "blindfold way," in: Forbes, *Travels Through the Alps*, p. 59.

157 Kane's letter to Agassiz: *Letter to Professor Agassiz of May 8, 1852*. In the library of the American Philosophical Society, Philadelphia.

159–160 Maclaren's idea about sea level is reported in: Cunningham, *James David Forbes*, p. 57. For a survey of Maclaren's life, see: "Charles Maclaren [obituary]." *Scotsman*, September 12, 1866.

163 For Murchison's attack on Agassiz's theory, see: Murchison, R. I. "On the Glacial Theory." *Edinburgh New Philosophical Journal* 33 (1842): 124–140. (This issue of the *Edinburgh New Philosophical Journal* has many important articles on glaciology.)

164–165 The story of von Buch at the 1845 meeting of the Swiss Society of Natural Sciences is presented in: Carozzi, "Agassiz's Amazing Geological Speculation," p. 79.

164 Humboldt quotes Madame de Sévigné in: ibid., p. 77.

165–166 Von Buch's strengths as a hiker are quoted in: ibid., p. 67.

Part V: Thrust Home

174 Kane on glacier's interior temperatures, from: Kane, *Arctic Explorations*, vol. 2, p. 208.

181–184 The story of Lyell, Agassiz, and Edward Forbes is not included in standard biographies (see Lurie, *Louis Agassiz*, p. 119, for the little he says of the Southampton meeting), but it is told in Lyell's letters: see letters dated September 26, 1846 (*Life, Letters, and Journals*, vol. 2, p. 104); October 14, 1846 (ibid., pp. 106–110); and especially July 29, 1849 (ibid., p. 155).

184 Agassiz on the line engraving of the glacier is quoted in: Lurie, *Louis Agassiz*, p. 77.

186 Agassiz on American look toward future quoted in: ibid., p. 124.

192 Kane on the "vast undulating plain," from: Kane, *Arctic Explorations*, vol. 2, p. 271.

194–199 The Lake Superior tour is described in: Agassiz, Louis, with a narrative of the tour by J. Elliot Cabot. *Lake*

Superior: Its Physical Character, Vegetation, and Animals, Compared with Those of Other and Similar Regions. Boston: Gould, Kendall & Lincoln, 1850.

199–200 Forbes's collapse is thoroughly explored in Cunningham, *James David Forbes*, which is also the source for information about Forbes's private recognition of the Ice Age evidence in north England and Scotland.

212 Agassiz's fiftieth birthday party is noted in several sources, including: Longfellow, Samuel, ed. *Life of Henry Wadsworth Longfellow, with Extracts from His Journals and Correspondence*. Boston: Houghton Mifflin Co., 1891; and Emerson, *Journals and Miscellaneous Notebooks*.

213 For the full text of Longfellow's birthday poem, see: Emerson, Edward Waldo. *The Early Years of the Saturday Club, 1855–1870*. Boston: Houghton Mifflin Co., 1918.

214 On Emerson speaking badly in impromptu contexts, see: Baker, Carlos. *Emerson Among the Eccentrics: A Group Portrait*. New York: Viking, 1996.

214 Emerson on "our thesis": *Emerson, Journals and Miscellaneous Notebooks*, p. 184.

215 William James says Agassiz trained whole generation of naturalists, in: James, William. *Memorial to Agassiz, Given on the 50th Anniversary of His Arrival in America*, Carbon copy of typescript, marked "Return to A. Agassiz," Museum of Comparative Zoology, Harvard University, Cambridge, Mass.

217 "…draw your rule from the genius of that which you do," from: Emerson, *Journals and Miscellaneous Notebooks*, p. 192.

217 Agassiz urged Longfellow to write, from: Longfellow, ed., *Life of Henry Wadsworth Longfellow*, vol. 3, p. 396.

217 Lyell's acceptance of glacier in Switzerland is told in: Lyell, *Life, Letters, and Journals*, vol. 2, pp. 245–280.

Part VI: Two Tragedies and a Triumph

228 The story of the Royal Geographic Society's meeting is found in: Corner, *Doctor Kane*, pp. 261–262.

230 George Templeton Strong quotation in Corner, *Doctor Kane*, p. 230.

233 For Lyell's "Greenlandization" of the Ice Age, see: *The Geological Evidences of the Antiquity of Man, with Remarks on Theories of the Origin of Species by Variation*. London: John Murray, 1863. The Greenland quotations at the opening are from pp. 243, 238, 237, 246, 290. Later quotation about uninhabited Scandinavia, p. 241

240 Agassiz's justification of nature studies, from: Agassiz, Louis. *The Structure of Animal Life: Six Lectures Delivered at the Brooklyn Academy of Music in January and February, 1862*. New York: C. Scribner & Co., 1866, p. 1.

241 Lyell says Agassiz has "gone wild" quoted in: Lurie, *Louis Agassiz*, p. 354.

241 Agassiz's personal response to Lyell, in: Agassiz, Louis. *Notes About Lyell's Post-Brazil Jest*. Undated, handwritten remarks on two sides of small page headed "Museum of Comparative Zoology."

242 "Don't you find a meaning…," from: Lurie, *Louis Agassiz*, p. 374.

245 "…there are now two of us who know," from: Petersen, Houston, ed. *Great Teachers*. New York: Vintage Books, 1946, p. 215.

Bibliography

Where to Begin

The ten works listed below will get a reader off to a good start in learning more about the discovery of the Ice Age and the main figures.

Agassiz, Louis. *Studies on Glaciers Preceded by the Discourse of Neuchâtel.* Edited and translated by Albert V. Carozzi. New York: Hafner Publishing Co., 1967.

Carozzi, Albert V. "Agassiz's Amazing Geological Speculation: The Ice Age." *Studies in Romanticism* 5, no. 2 (winter 1966): 57–83.

Corner, George W. *Doctor Kane of the Arctic Seas.* Philadelphia: Temple University Press, 1972.

Cunningham, Frank F. *James David Forbes, Pioneer Scottish Glaciologist.* Edinburgh: Scottish Academic Press, 1990.

Forbes, James D. *Travels Through the Alps of Savoy and Other Parts of the Pennine Chain, with Observations on the Phenomena of Glaciers.* Edinburgh: Adam & Charles Black, 1843.

Kane, Elisha Kent. *Arctic Explorations in the Years 1853, '54, '55.* 2 vols. Philadelphia: Childs & Peterson, 1856.

Lurie, Edward. *Louis Agassiz: A Life in Science.* Paperback edition, with additions. Baltimore: Johns Hopkins University Press, 1988.

Lyell, Charles. *Life, Letters, and Journals.* Edited by his sister-in-law Katherine M. Lyell. 2 vols. London: John Murray, 1881.

______. *Principles of Geology.* 3 vols. Chicago: University of Chicago Press, 1990.

Wilson, Leonard G. *Charles Lyell: The Years to 1841: The Revolution in Geology.* New Haven, Conn.: Yale University Press, 1972.